Sustainable Production and Food Security

An Overview through Climate Smart Agricultural Interventions

Sustainable Production and Food Security

An Overview through Climate Smart Agricultural Interventions

Edited by

Ratnesh Kumar Jha
Dr. Rajendra Prasad Central Agricultural University, India

Abdus Sattar
Dr. Rajendra Prasad Central Agricultural University, India

Asim Biswas
University of Guelph, Canada

Latief Ahmed
SKUAST, India

Sudeshna Das
Bihar Agricultural University, India

Pawan Kumar Srivastava
Dr. Rajendra Prasad Central Agricultural University, India

World Scientific

NEW JERSEY · LONDON · SINGAPORE · BEIJING · SHANGHAI · TAIPEI · CHENNAI

Published by

World Scientific Publishing Co. Pte. Ltd.

5 Toh Tuck Link, Singapore 596224

USA office: 27 Warren Street, Suite 401-402, Hackensack, NJ 07601

UK office: 57 Shelton Street, Covent Garden, London WC2H 9HE

Library of Congress Cataloging-in-Publication Data

Title: Sustainable production and food security : an overview through climate smart
    agricultural interventions / edited by Ratnesh Kumar Jha, Abdus Sattar, Latief Ahmed,
    Asim Biswas, Sudeshna Das, Pawan Kumar Srivastava.
Description: Hackensack, NJ : World Scientific, 2025. | Includes bibliographical references
    and index.
Identifiers: LCCN 2024028343 | ISBN 9789811296079 ebook other |
    ISBN 9789811296062 ebook | ISBN 9789811296055 hardcover
Subjects: LCSH: Sustainable agriculture | Agricultural systems |
    Agricultural ecology | Crops and climate
Classification: LCC S494.5.S86 S8935 2025 | DDC 338.1--dc23/eng/20241118
LC record available at https://lccn.loc.gov/20240283433

British Library Cataloguing-in-Publication Data
A catalogue record for this book is available from the British Library.

For any available supplementary material, please visit
https://www.worldscientific.com/worldscibooks/10.1142/13926#t=suppl

Desk Editors: Aanand Jayaraman/Joy Quek

Typeset by Stallion Press
Email: enquiries@stallionpress.com

Preface

In an era marked by unprecedented challenges and trials, the intersection of sustainable production and food security has never been more critical. Among the most pressing threats to these systems is climate change, which is reshaping weather patterns, increasing the frequency of extreme events, and altering the very conditions under which we grow food. As the global population continues to upsurge, the food requirement intensifies, placing enormous pressure on our agricultural systems, natural resources, and the environment.

The urgent need for innovative solutions has never been clearer. We must explore sustainable practices that not only adapt to the realities of a changing climate but also mitigate its effects. This book, *Sustainable Production and Food Security*, aims to examine these critical connections and provide a roadmap for integrating climate resilience into agricultural practices for a resilient food future. This book brings together insights from diverse fields, including integrated agriculture, agrometeorology, environmental science, and artificial intelligence, to explore how sustainable practices and meticulous selection and adoption of crops and their cultivation can enhance food security while ensuring the health of our planet. Throughout the chapters, we delve into key concepts such as regenerative agriculture, agroecology, and the role of technology in enhancing sustainability. Each section is designed to provoke thought, inspire change, and encourage a deeper understanding of the complexities involved in feeding a growing world without compromising the health of future generations.

As you journey through this book, we invite you to consider your role in this global narrative. Whether you are a farmer, a policy-maker, a student, or a concerned citizen, your actions and decisions can contribute to a more sustainable and equitable food system. Together, we can reimagine production practices that nourish both people and the planet, paving the way for a brighter, more sustainable future inspiring one another to foster change in our communities and beyond.

About the Editors

Ratnesh Kumar Jha is a Professor in the Department of Agronomy, Dr. Rajendra Prasad Central Agricultural University Pusa, Bihar, India. Presently, he is heading the Center for Advanced Studies on Climate Change as Project Director. He is the Principal Investigator of project "Scaling up Climate Smart Agriculture (CSA) through Mainstreaming Climate Smart Villages (CSVs) in Bihar" funded under the National Adaptation Fund for Climate Change, Ministry of Environment, Forests and Climate Change, Government of India. Under the leadership of Jha, the Centre is working in collaboration with different national and international organizations such as CPRI, Shimla for Quality Potato Seed Production through aeroponics and ARC, National Institute of Disaster Management (NIDM), Borlaug Institute for South Asia — International Centre for Maize and Wheat Research (BISA–CIMMYT), and International Rice Research Institute (IRRI) as well as International Research Center for Semi-Arid and Tropics (ICRISAT). He has more than 50 research papers in high NAAS-rated journals, 10 books, and has participated and presented in more than 100 seminars, symposia, and conferences, as well as published more than 12 training manuals. Jha is a Nodal Officer of National Agriculture Disaster Management Plan. He has guided more than 10 M.Sc. Ag and Ph.D. students as major and co-advisor.

Abdus Sattar is an Associate Professor-cum-Senior Scientist (Agrometeorology) and Principal Investigator, All India Coordinated Research Project on Agrometeorology, RPCAU, Pusa (Samastipur) Bihar. As Principal Investigator, he is also handling other agrometeorological projects, *viz.*, National Innovations on Climate Resilient Agriculture (NICRA) and Gramin Krishi MausamSewa (GKMS). He has also handled a project funded by Space Application Centre (SAC), ISRO, Ahmedabad. He has published many research papers in peer-reviewed national and international journals, technical bulletins, and books. Sattar has contributed significantly to agrometeorological research and enhanced farm productivity and farmers' income through climate services and risk management. He is credited with modeling rice-wheat cropping system in middle Indo-Gangetic plains and has developed the climate smart irrigation software on rice. In recognition of his contributions, Rajendra Agricultural University, Pusa, Bihar (Now Dr. RPCAU, Pusa) gave him the "Best Teacher/Scientist Award" in 2011. He received the "Best Ph.D Thesis Award" from the Association of Agrometeorologists in India. He is also the recipient of the "Best Research Paper Awards" given by the university and the Association of Agrometeorologists.

Asim Biswas is a Professor, Canada Research Chair (Tier 1) in Digital Agriculture and the OAC Research Chair in Soils and Precision Agriculture at the School of Environmental Science, University of Guelph, Canada, and a member of the Royal Society of Canada College. His research program on sustainable soil management is focused on increasing the productivity and resilience of our land-based agri-food production systems in an environmentally sustainable way while accounting for changing climate, economy, and production methodologies. Currently, he runs a research program on sustainable soil management funded by federal and provincial bodies as well as industries, grower's associations, and international organizations. He has authored and coauthored 250+ peer-reviewed journal papers, 225+ conference abstracts, 10 edited and authored books, 23 book chapters, delivered radio and TV interviews, and was granted a patent. He was invited to deliver Plenary and Keynote talks (60+) around the world and currently teaches multiple undergraduate, graduate, and special courses. Currently, he is an Associate Editor for several journals and a Guest Editor for a series of special issues. He is a past President of the Canadian Society of Soil Science (CSSS), member of the Executive council of the Royal Society of Canada (RSC) College of New Scholars, Canadian Lead to the International Society of Precision Agriculture (ISPA), Chair of the Spatial Statistics and On-Farm Experimentation community of the American Society of Agronomy (ASA), Chair of the Proximal Soil Sensing Working Group of the International Union of Soil Science (IUSS), and a part Chair of the Soil Physics and Hydrology Division of the SSSA. He is a member of 17 professional societies and leads various committees in those societies.

Latief Ahmad is a Scientist (Sr. Scale) at Dryland Agricultural Research Station, Rangreth Sher-e-Kashmir University of Agricultural Sciences and Technology, Shalimar Kashmir, J&K India. He obtained his Bachelor's degree in Agriculture from the Faculty of Agricultural Sciences, CCS-SDS College, Aligarh, and his Master's and Doctoral Degree in Agronomy from the School of Agriculture, Allahabad Agricultural Institute, Naini Allahabad U.P., India. The author has published more than 90 research articles in peer-reviewed international and national journals He has authored 4 books, 12 book chapters, and 24 extension bulletins. He has attended more than 40 national and international conferences/Training programs, handled 5 national and international projects, and is a life member of 7 scientific associations. He has received the best research proposal award by SERB-DST. His areas of interest include agrometeorology, climate change studies, crop simulation modeling, precision farming, CSA, water management, and crop science. Ahmad is also a Visiting Scientist at the University of Guelph, Ontario, Canada.

Sudeshna Das is an Assistant Professor, Plant Physiology at Bihar Agricultural University, Sabour, Bihar, India. Her field of work encompasses monitoring climate change on crop production and conducting on-field and controlled facility experiments to understand the effect of drought and heat stress in crop plants in order to screen for drought and heat tolerant and opportunistic varieties. She also conducted experiments to study ameliorative impacts of Silicon under drought conditions. She has authored multiple research papers and book chapters in peer-reviewed international journals pertaining to stress physiology and climate change.

Pawan Kumar Srivastwa is a Research Associate at Climate Resilient Agriculture Programme. He did his M.Sc. in Botany from Babasaheb Bhimrao Ambedkar University Muzaffarpur and Ph.D. in Botany from Jai Prakash University, Chapra. As Research Associate, he has worked on a number of projects from 2011 to till now, *viz.*, National Agricultural Innovation Project C-3, Sustainable and Resilient Farming Systems Intensification (SRFSI) CIMMYT funded project, and Climate Resilient Agricultural Project. He has also published several research papers.

Contents

Chapter 1

Integrated Farming Systems for Flood-Prone Ecosystems

K.G. Mandal

ICAR-Mahatma Gandhi Integrated Farming Research Institute (MGIFRI), Piprakothi, Motihari, Bihar
kgmandal@gmail.com

Abstract

Flood-prone ecosystems are mostly productive, but the productivity is not exploited to its full potential because of the associated problems of seasonal and permanent waterlogging. The occurrence of cyclic anaerobic and anaerobic soil conditions renders soils of the region not very useful, whereas soils have high nutrient status. Considering the demand for more food, and nutrient security for human beings currently and in the future, it would be essential to manage those challenged ecosystems, especially their natural resource base, i.e., the soil, crop, and vegetation. In this chapter, attempts have been made to elucidate the extent of food insecurity in different regions of the world, natural resource base of the ecosystem, extent of flood-prone areas and waterlogging in the country, how is variation of the integrated farming systems (IFS) over time and space, interaction of different components of IFS and their relationship with different situations, integration with rice-fish co-culture, aquatic crops, and different other potential IFS, which would increase the productivity of both water and soil. The whole perspective of IFS has been discussed addressing flood-prone and waterlogged conditions, and the targets of SDGs. This is relevant for actual farming situations in many parts of India and other Asian countries.

Keywords: Diversification, Integrated farming, Food security, Waterlogged Ecosystems.

Introduction

The United Nations has set 17 sustainable development goals (SDGs). SDG 2.0 is "End hunger, achieve food security and improved nutrition and promote sustainable agriculture" by 2030. About 768 million people are still undernourished around the world, that is those who are food insecure. This has been reported by international organizations, *viz.*, FAO, IFAD, UNICEF, WFP, and WHO (2021). Out of the 768 million, more than half of the undernourished live in Asia, and more than one-third in Africa. In India, about 15.3% are undernourished. Therefore, agricultural systems need to be strengthened; at the same time, natural resource management is facing several challenges such as the degradation of land, water-related issues like both excess and shortage of water, climate change, etc. The base of different production systems needs to be sustainable. There are challenges for flood-prone and waterlogged ecosystems. It is required to bring forward suitable integrated farming systems befitting to water-congested ecosystems (Mandal *et al.*, 2021).

Food security was defined as a state or condition "when all people, at all times, have physical, economic and social access to sufficient, safe and nutritious food to meet their dietary needs and food preferences for an active and healthy life" during the World Food Summit in 1996 (FAO, 1996), and "The State of Food Insecurity in the World", which was prepared jointly by the FAO, WFP, and IFAD (2012) focused on the necessity of economic growth to overcome malnutrition, poverty, and hunger. Again, it has been reported recently by international organizations that the food insecurity percentage has increased (2014–2016 to 2018–2020) under the "severe" and "moderate to severe", based on the food insecurity experience scale, and the increase is prominent in Asia, Africa, and Latin America and the Caribbean (Fig. 1). Although the prevalence of undernourishment (PoU) remained unchanged virtually during 2014–2019 around the world, the PoU increased from 8.4% to 9.9% between 2019 and 2020 due to COVID-19 pandemic (FAO, IFAD, UNICEF, WFP, and WHO, 2021). The largest undernourished population belongs to Asia. In 2020, out of 768 million undernourished, more than half, i.e., 418 million are in Asia, and more than one-third, i.e., 282 million in Africa. In India, about 15.3% of the population is undernourished. As the trend continues, questions and doubts are arising whether

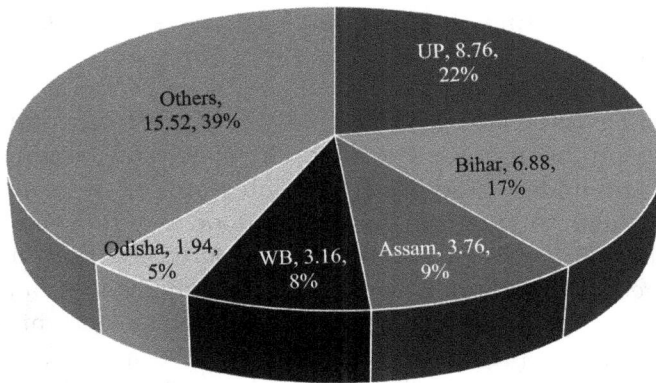

Fig. 2. Major flood-prone areas in Indian states with area and percent of the total flood-prone area.

Bihar witnesses widespread damage due to riverine flood. The degree of the riverine flood is increased due to continuous rainfall and with high intensity in the river catchment areas. Soil water saturation is continuous for a long period. Coastal flooding (i.e., storm surge) is also a frequent occurrence in coastal areas of India. Vast stretches of the coastal area get inundated by seawater intrusion due to intense windstorms, high tides, and different intensities of cyclonic events. On average, floods of different types cause waterlogging (perennial and seasonal) in about 11.6 million ha in the country. The low-lying areas are mostly affected. These waterlogging ecosystems may be considered the most challenging with respect to the management of natural resources, especially water.

In eastern India, the problem of waterlogging is due to all types of floods — riverine, cyclone-associated floods, and flash floods. In Bihar, which is situated in the middle Gangetic plains, the waterlogging problem is the most serious due to flooding from Nepal-based rivers, *viz.*, Gandak, Burhi Gandak, Bagmati, Koshi, Kamla-Balan, among others (~0.8 million ha every year). Rivers carry a high volume of discharges and sediments from the Himalayas. South Bihar also is heavily affected by excess discharges in the Ganga tributaries (Bhanumurthy *et al.*, 2010). In Bihar, there are about 57% population who are affected by flood, and most of them belong to north Bihar. In Bihar, 28 districts out of 38 are flood-prone, of which 15 of them are most vulnerable (Sinha, 2008, 2009; Sinha *et al.*, 2008).

Integrated Farming Systems to Address Waterlogging Ecosystems

Waterlogging does not possess a very congenial environment for growing many field crops. However, any successful farming system interventions under these conditions integrate the field crops as a key component to cater to the diverse food requirements of farm households, with diversified agriculture activities that may help farmers generate additional income, more employment, and better livelihoods by following the effective resource management and sustainable systems, preserving the resource base and ensuring high environmental quality. Ideally, there are several components of integrated farming systems, *viz.*, crop component (field crops, agroforestry, fodder crops, horticultural and plantation, aquatic crops, etc.), animal component (dairy, poultry, and duck rearing), fisheries and aquaculture, mushroom cultivation, beekeeping, sericulture, etc.). The farmer's realization is the physical output from the system components, economic returns, employment, and of course, the livelihood improvement (Fig. 3, Mandal *et al.*, 2021). The

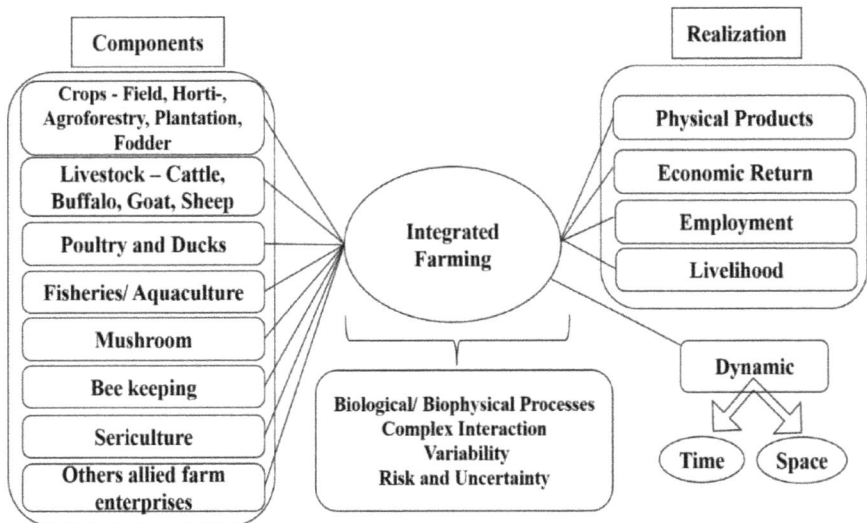

Fig. 3. Schematic diagram indicating integrated farming system components and realizations from the system.
Source: Mandal *et al.* (2021).

systems approach in integrated farming involves complex interactions among different components as there exist biological and/or biophysical processes. The integrated farming system is dynamic in the time and space dimensions; risks and uncertainty are also associated.

Variations in the natural resource base, i.e., land situations, water resources — its excess and limited availability during post-rainy seasons, resource endowments available with farmers, nutrient flow in the agroecosystems and soil sickness, water pollution, climatic variability, and extremes, group farming or partnership approach, government incentives and schemes, market intelligence and farmer's own choice of crops and enterprise development, etc., govern the actual farming situations to be adopted in a particular location or site. In that sense, an integrated farming system is highly site-specific and situation-specific. At the same time, the demand or the requirements from the systems are multiple, although interlinked; for example, maximizing yield is the target along with maximizing profit, with the aim to improve livelihood, nutrition, self-sustenance, employment generation, and environment protection (Fig. 4). These variable situations and requirements from the systems have made the actual farming systems dynamic (Mandal *et al.*, 2021). To address the waterlogging situations, there are several potential approaches in integrated farming systems, as discussed in the following section, which would fulfill the requirements from the systems and ensure food and nutrition security of farmers, at least partially. A diagram of the integrated farming system adopted in Odisha is presented in Fig. 5.

In case of severe waterlogging, it becomes difficult to grow any other crop except rice during the *kharif* season. Also, after the monsoon, due to the rise in the groundwater table in many areas, other cash or remunerative crops are not very suitable. Overall, the rice-based farming systems have been found immensely beneficial with some land modification technologies combined with effective water management, such as multiple use of water, together with soil reclamation and crop management to grow certain crops after the monsoon season or receding of floodwater. By this approach, rice-based farming systems integrated with multi-enterprise can contribute toward production sustainability and nutrition security for the farmers.

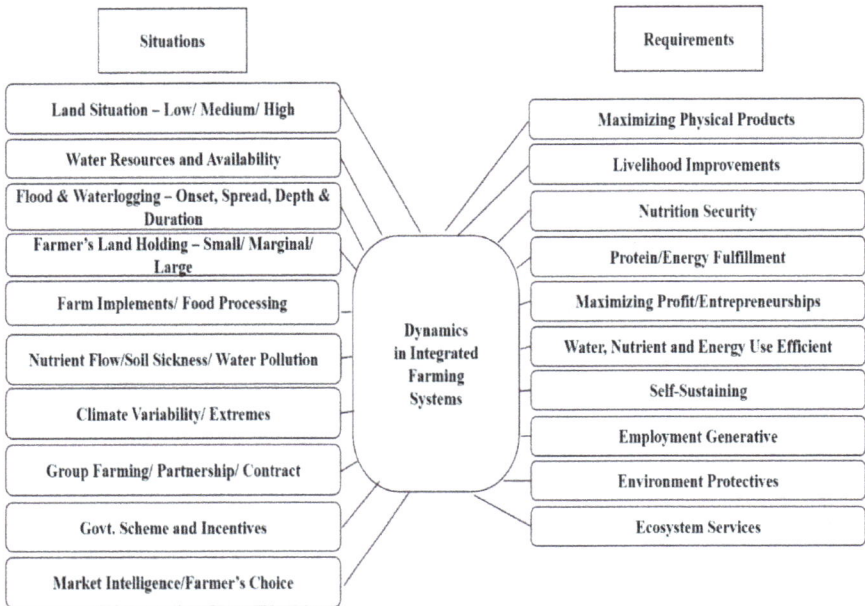

Fig. 4. Schematic diagram of an integrated farming system with variable situations and demand or requirements from the system components and realizations from the system.

Source: Mandal *et al.* (2021).

Fig. 5. Integrated farming system at Dasphala, Nayagarh, Odisha.

Rice-Fish Systems

The practice of rice-fish integrated farming in India dates back almost 1,500 years. An integrated system of rice and fish culture is predominant in states, *viz.*, Assam, Tripura, Andhra Pradesh, Bihar, Kerala, West Bengal, and Tamil Nadu, but further improvement would be required. There is comparatively plenty of water in eastern India during the monsoon months (June–September), and rice is the only field crop that is grown in those areas. Other aerobic crops are suitable. Integration of fish with rice proves beneficial with some minor interventions like the creation of refuges in the field, where fish will take shelter and grow apart from feeding in the rice fields. This is a kind of micro watershed in the rice field; fish refuge is generally made in 10–20% of the area, and the raising of strong and wide dykes all around the field protect the micro-environment and accommodate the integration of different components like fruit crops, vegetables, floriculture, livestock, birds, mushroom, agroforestry, apiculture, etc. It is required to increase the productivity of both water and land; *in-situ* and *ex-situ* conservation of rainwater and integration of fish/prawn culture are found to be a viable and easily adaptable technology for small and medium farmers; the system can be an option for business development. Evidence show that in irrigated rice ecosystems, a refuge area of 9% of the rice field resulted in the highest return without using any pesticide (Mohanty and Mishra, 2003).

There is huge potential for rice-fish farming in north Bihar which is rich in water resources comprising rivers, *chaurs*, ox-bow lakes or *mauns*, reservoirs, ponds, tanks, etc. The total fish production of Bihar has reached 6.4 lakh tons in 2019–2020, but not sufficient enough to meet the requirement. Fish productivity of floodplain wetlands of Bihar is reported to be only 50–220 kg/ha/year, which can be increased several folds through scientific interventions (Ayyappan *et al.*, 2017). The most preferred fish species under rice-fish integration are *Catla catla*, *Cyprinus carpio*, *Cirrhina mrigala*, *Labeo rohita*, *C. mrigala*, *Puntius gonionatus*, and *Macrobrachium rosenbergii* (Mohanty *et al.*, 2004).

Rice–fish interactions in integrated farming create an environment of mutualism and synergism where both rice and fish benefit from each other and ultimately render the rice field more productive, profitable, as well as environmentally sustainable. Rice-fish farming in rainfed lowland increases effective tillers by 9–14% and panicle weight by 5–17% in rice; enhances soil N uptake by 10–11.8% and Fe uptake in straw by two-folds over rice alone (Sinhababu *et al.*, 1983, 1992; Panda *et al.*, 1987). Continuous addition of fish excreta in the rice-fish system increases the soil organic carbon content by 7% and exchangeable NH_4-N by 25%, available-P by 6% with an increase in grain yield rice by about 5–15% and straw yield by 5–9% under rainfed lowland ecologies (Sinhababu et al., 1998). There are other concomitant benefits of rice-fish culture. Fishes act as effective bio-control agents for major rice pests like hoppers, stem borer, gall midge, leaf folder, and snails (Sinhababu and Majumdar, 1981). Insecticides though are not harmful to the reared fishes instantly show varying degrees of deleterious effects on growth, survival, and yield (Sinhababu and Rajamani, 2000). Fish culture in rice fields controls weeds directly by feeding, mechanical injury, and constant flooding (Patra and Sinhababu, 1995). This system improves soil fertility through increasing N and P availability (Dugan et al., 2006). Fish increases aeration in rice fields while scavenging foods and helps control aquatic weeds, algae, flies, snails, and insects, and thus rice-fish farming is considered an important component of integrated pest management in rice. By controlling insect pests and diseases, the integration of fish with rice lowers the cost of rice production. Rice fields, on the other hand, provide planktonic, epiphytic, and benthic food for fish and also maintain favorable water temperature for fish as rice plants provide shade during the hot days.

Pond refuge connected with two side trenches was developed to increase farm productivity and income from rainfed waterlogged lowland areas in Cuttack (Annie Poonam et al., 2019). Field design included wide bunds (dykes) all around, a pond refuge connected with trenches on two sides and a guarded outlet, and with the construction of a small low-cost duck house, poultry unit, and goat shed on bunds; suitable rice varieties, such as Gayatri, Sarala, CR Dhan 500, CR Dhan 505, Jalmani, and Varshadhan could be grown with stocking of the mixed fish culture of 3″–4″ size and prawn juveniles of 2″–3″ size in a ratio of 3:1. After rice, various crops like watermelon, green gram, sunflower, groundnut, sesame, and vegetables could be grown in the field with limited irrigations from the harvested rainwater. On bunds, different crops/plants were grown including vegetables, spices, and pineapples in shades, fruit crops (dwarf papaya, banana, coconut, and areca nut), flowers (tuberose and marigold), straw and oyster mushroom cultivation, bee rearing (2–3 bee boxes), and short-duration agroforestry (Acacia spp.).

The rice-fish farming system can annually produce around 16–18 t of food crops, 0.6 t of fish and prawn, 0.55 t of meat, 8,000–12,000 eggs besides flowers, fuelwood, and animal feed as rice straw and other crop residues from 1 ha of a farm. The net income in the system is about Rs. 76,000 in the first year. Subsequently, it increases to around 1,30,000 in the sixth year. This system thus increases farm productivity by about 15 times and net income by 20 folds over the traditional rice farming in rainfed lowlands. It also generates additional farm employment of around 250–300 man-days/ha/year. Rice-fish-horticulture model having pond refuge connected with two side trenches and wide dykes all around for rainfed lowland rice ecology developed at Regional Rainfed Lowland Rice Research Station (NRRI-RRLRRS), Gerua, Assam produced 5–5.5 and 3.5–4 t ha^{-1} of rice in *sali* and *ahu* season, respectively. After establishment years, the system produced REY of 9.05 t/ha and generated an average net income of Rs. 52,380 per annum from a 0.5 ha area with employment generation of 206 man-days/ha/year. With respect to environmental aspects, of course, previous research shows that emission of methane (CH_4) is increased with fish in the rice field but nitrous oxide (N_2O) emission was found to be relatively low during the entire cropping period except toward maturity when the water recedes leaving the field dry (Datta *et al.*, 2009; Bhattacharyya *et al.*, 2013).

Rice–Fish–Duck/Poultry Farming

The rice–fish–duck farming provides maximum synergy and utilization of available resources benefitting small and marginal farmers, especially, in tribal-dominated areas under medium-deep or deep-water lowland rice ecosystems. Ducks and fishes control the harmful insects and weeds, droppings are utilized as organic manure, activities, *viz.*, continuous movement, scooping, and churning of soil aerate the soil-water environment which increases the availability of nutrients to the rice crops. It also reduces the cost of cultivation and increases productivity by providing sustainability, financial stability, and employment to the farm families. The rice–fish–duck integration in rice-rice farming systems annually produces about 9–10 t food crops, 0.7 t fish prawn, 0.5–0.7 t of meat, and 25,000 eggs per ha. The net income is around Rs. 1,10,000–1,30,000 depending upon the system components with a benefit-cost ratio of 2.5–2.8 (Nayak *et al.*, 2021). Rice–fish–poultry farming system has been demonstrated on 430 farm holdings in 12 villages of Cuddalore, Villupuram, Nagapattinam, and Thiruvannamalai districts of Tamil Nadu by the Annamalai University, with transplanted rice in 200 m^2

Fig. 6. Integrated farming system with fish and poultry.

area, 20 poultry birds kept in cages of size $1.8\,\text{m} \times 1.2\,\text{m} \times 1\,\text{m}$ and 100 fingerlings (*Rohu, Mrigal, Catla*) in trenches of $20\,\text{m}^2$ area in each unit. There was an annual increase in net return by Rs. 33,000–50,500 per ha for two and three crops, respectively. The addition of poultry manure due to poultry dropping was 11.4–19.6 t/ha and also pest suppression by 17–27% (Shrivastava, 2018).

Pond management with fish, poultry, and vegetables has proved to be an excellent approach for sustainable production, income, and employment generation for the resource-poor rural households in Ethiopia (Lemma, 2013). The poultry-fish farming system can be integrated to reduce the cost of fertilizers and feeds in fish farming. It can be easily maintained in small areas and produce the product in a short time. The addition of organic fertilizers like poultry litter to a fish pond, that is integrating poultry farm with fish, increases the water nutrients for better production of fish by resolving the problem of fish feed. A poultry house can be constructed from locally available materials after the excavation of a fish pond. The walls of the house may be made from eucalyptus wood and plastered with mud. The roof of a poultry house can be constructed with thatch, plastic sheet, or tin (Lemma, 2021). An example of integrated farming system with fish and poultry at ICAR-MGFRI, Motihari, Bihar is presented in Fig. 6.

Makhana Crop with Fish Culture/Water Chestnut and Rice

Makhana (*Euryale ferox*) and water chestnut (*Trapa bispinosa, T. natans*) cultivation are popular among the fisherman communities in waterlogged ecosystems of north Bihar, especially in the

Darbhanga district. These aqua crops are cash crops and provide nutritional and livelihood security to a considerable section of the population in the region. *Makhana* and water chestnut are widely utilized in India as non-cereal diets and have great demand during festivals and ritualistic fasts. In lowland areas of Darbhanga district in Bihar where cultivation of other crops is not possible due to waterlogging, integration of *makhana* with fish and water chestnut, and field-based system of *makhana* cultivation is found to be beneficial (ICAR-RCER, 2014); integration of fish and water chestnut with *makhana* was developed in 50 ha of land in Darbhanga district. The net economic benefit was Rs. 52,435 as compared to Rs. 20,614 in the traditional system of *makhana* alone. Kumar *et al.* (2012) have suggested integrated farming systems involving crop or crop sequence, livestock, fish, and horticulture for marginal and small, and medium and large farmers in the waterlogged areas in Bihar.

Remunerative Crops after Rice in Coastal Waterlogged Ecosystems

After harvesting rice, it is possible to grow pulses, oilseeds, vegetables, and other crops during winter/summer seasons with proper water management in a particular farming system. Crops like chili, sunflower, groundnut, lady's finger, watermelon, and basella can be grown after rice during the dry season (Saha *et al.*, 2007). The tropical tuber crops such as sweet potato and yams are well adapted to coastal agroecosystems. Sweet potato can be grown in lowland rice fallow in a deepwater rice-fish system with the variety "Kishan" under conventional tillage and "Sourin" under minimum tillage conditions realizing root yield of 20.98 and 13.95 t/ha, respectively (Nedunchezhiyan *et al.*, 2013). Other crops, *viz.*, pulses like black gram, green gram, etc., or oilseeds, *viz.*, sunflower, groundnut, sesamum, etc., can be grown during the dry season (January–early April) with limited irrigation from harvested water (Saha and Biswal, 2004). In different on-farm IFS models in coastal saline areas, watermelon, chili, and sunflower were found to be the most promising crops for both medium and high salinity areas, while lady's finger was suitable and the most remunerative under low to medium salinity conditions (Singh *et al.*, 2006).

Horticulture-Based Integrated Farming Systems

The cultivation of horticultural crops under a horti-based farming system plays an important role in the farming systems, which increase nutritional security, employment, and sustainable development in any agroecosystems, and waterlogged ecosystems are not an exception. Various problems that small and marginal farmers face are solved by adopting horticulture-based integrated farming systems (Kumar *et al.*, 2012). Surface drainage by land modification technology and recycling of drainage water seem to be feasible in waterlogged ecosystems. Rainwater harvesting and its recycling can increase productivity and diversify agricultural systems including remunerative agricultural and horticultural crops and pisciculture in an integrated manner (Das *et al.*, 2014a). In some cases, under waterlogged areas, despite surface drainage, excess water cannot be drained out due to land topography. The groundwater level of those lands is so near to the surface that water cannot be drained out to the main drains. In those cases, restoration of seasonally waterlogged lands is possible through the integration of various techniques of land shaping, i.e., pond-cum-raised bed system and pond system. The ratio of pond-cum-raised bed system depends on location and landscape. Land shaping technique includes modification of the land surface primarily for harvesting of rainwater and creating a source of water for irrigation, reducing drainage congestion, growing of multiple and diversified crops all year round, and efficient use of stored water. Through these land modification techniques, there is ample scope for growing fruit and seasonal vegetable crops on the raised beds and the pond dykes, and fish culture in the sunken water body of different depths and in ponds. This approach and practice would ensure nutrition security for farmers, employment generation throughout the year, and enhance the productivity of the whole land-water water. A multi-enterprise integrated farming (apiculture, mushroom cultivation, vermicomposting, medicinal and aromatic crops) would also be possible in waterlogged areas. Various studies have shown that sustainable development is possible due to horticulture-based integrated farming systems.

In waterlogged ecosystems, popular models are mainly pond-dyke system, fish-rice-duck/poultry-vegetable, fish-water chestnut/makhana-vegetables, and fish-cattle/pig-duck/poultry-vegetable. An on-farm study in Khurda district of Odisha showed that pond-cum-raised system fetched 20–40% more yield than the conventional lowland cultivation system of different crops, and the B/C ratio of 2.45 in the improved system (Sahoo *et al*, 2006). Similarly, a horti-based integrated farming system (Patel *et al*., 2020) comprising four pond-dyke models (model I: two rows of litchi and banana + seasonal crops, model II: two rows of litchi and papaya + seasonal crops, model III: two rows of litchi + banana in between two litchi plants + seasonal crops, model IV: two-row of litchi + papaya in between two litchi plants + seasonal crops) along with traditional cropping system practiced in low-lying areas, i.e., fallow-mustard-moong to compare with different models found the highest sustainable value index, total system productivity, and economic efficiency (0.78, 25.49, and 116.15) in model I, respectively, as compared to crop-based traditional systems. Five vegetable-based cropping sequences on raised beds and six rice-based sequences on sunken beds were tested and compared with rice monocropping under an organic production system. On raised beds, tomato–okra–french bean gave the highest rice equivalent yield (REY; 44.7 t ha^{-1}) followed by carrot–okra–french bean (42.5 t ha^{-1}). Rice (cv. Shahsarang 1)-pea (cv. Prakash) gave the highest REY (17.3 t ha^{-1}) on sunken beds (Das *et al*., 2014b). Plantation of litchi on ridges is recommended in low-lying areas to prevent waterlogging in the Philippines (Sotto, 2002). The overall design of litchi orchards should be established according to the topography. The raised-bed model in the lowlands and contour-making in steep land are the two main designs practiced in Vietnam (Hai and Van Dung, 2002). Selected case studies of family farming models and horticulture-based farming systems are reported for their role in farming systems suited to local climatic needs particularly suitable for waterlogged ecosystems.

Livestock-Based Integrated Farming in Waterlogged Ecosystems

Livestock is an important component of integrated farming, and is considered in some sites central to the livelihood of farmers. This system is a traditional practice in rural India among small and marginal farmers. Both income and employment generation

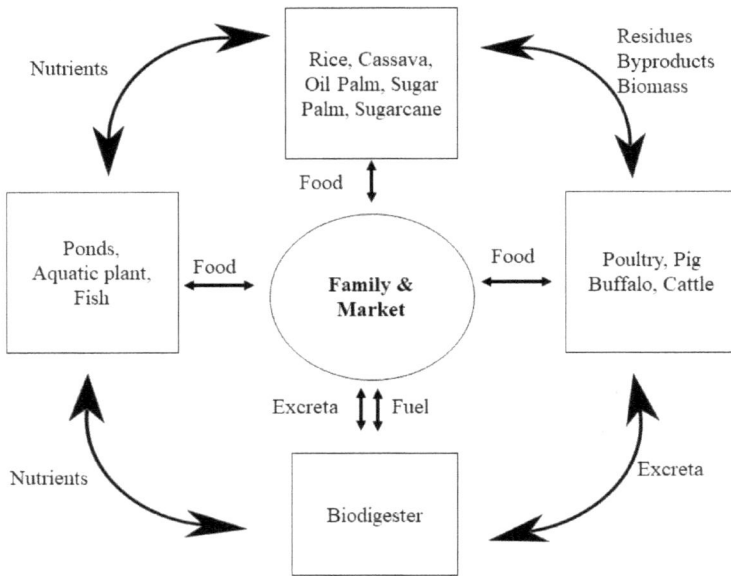

Fig. 7. Resource recycling of farm resources or wastes in Cambodia.
Source: Preston (2002).

are greatly associated with livestock-based farming. Livestock is an important natural capital asset for the poor, which they use to maintain a livelihood in times of crisis. In Cambodia, major emphasis is given to the productive cycling of farm resources or wastes, as exemplified by Preston (2002, Fig. 7), where importance is given to animals and recycling of byproducts, nutrients, and excreta. The use of local resources should be prioritized. Different components of the farming system work together in an integrated farming system resulting in higher total productivity. The output from one enterprise, thus improving the resource use efficiency. Due to fragmentation in the landholding of farmers, it is necessary to integrate land-based enterprises like fishery, poultry, and horticultural crops within the bio-physical and socio-economic conditions of the farmers to make farming more profitable and dependable (Behera *et al.*, 2004). India has huge diversified livestock resources enriching the lives of millions of people in rural areas.

Raising water table and waterlogged ecosystems, only crop-based agricultural outputs in an isolated system, cannot fulfill the daily needs of human diet and nutrition due to the decreasing trend of landholding per farm family and hence the diversification of crop-based agriculture with dairy, goat rearing, fishery, poultry, etc., is necessary for increasing the income of farmers. Due to the uncertainty of monsoon, the farmers are forced to adopt a judicious mix of agricultural enterprises like dairy, poultry, fishery, etc., which would suit their local situations and socioeconomic conditions. Integrated farming of fish and livestock consists of the culture of fish (or shrimp) associated with the husbandry of domesticated animals such as cattle, pigs, ducks, chickens, etc. This can be well-suited to the waterlogged ecosystems, especially in flood-prone areas like northern Bihar, parts of West Bengal, Odisha, etc. The highest production so far in integrated farming is with pigs, ducks, and chickens. In some areas, farmers also integrate geese, rabbits, goats, sheep, cattle, and water buffalo along with fish culture on a small scale (Khan *et al.*, 2014). The main fish species stocked in animal-fish pond systems, either in mono- or polyculture, are the common carp and some exotic varieties.

Livestock-crop-fish farming systems are particularly important for increasing the return from a limited land area and reducing risk by diversifying crops (Korikantimath *et al.*, 2008). The nutrient content in livestock waste will help the growth of phytoplankton and zooplankton in fish ponds. The byproducts of livestock such as manure, urine, and spilled feed can be used for aquaculture directly or indirectly based on the need and convenience. An adult cow produces approximately 4,000–5,000 kg of dung and 3,500–4,000 liter urines annually. For a pond size of one hectare, 5–6 adult cattle would be sufficient to provide adequate manure. In addition, 9,000 liters of milk and about 3,000–4,000 kg fish/ha/year can be produced in this system. The goat could be one of the best livestock to start with an integrated farming system for small and marginal farmers. Goats can be easily maintained in small areas or even in backyard systems without much investment compared to large animals like cattle and buffaloes. Goats can be easily housed on a raised platform above the pond or even in the embankment of the pond area.

Thus, it is very essential that an integrated farming system be developed, which will be a viable solution to increase agricultural production, food security, improved nutrition, and promote sustainable agriculture. This would help in achieving the UN's SDG No. 2

by the year 2030. An integrated farming system is also a promising enterprise development option for marginal and small farmers. For integrated farming systems, the underlying science is mostly known, but it is required to develop site-specific dynamic technology that will fit into variable situations. As the actual farm situations vary over time and space, variability should be assessed before recommendation. Of course, infrastructure development will be required concomitantly; if possible, it should be low-cost technology and local resource-based for the ease of adoption by resource-poor farmers, especially in waterlogged areas.

References

Annie, P., Saha, S., Nayak, P.K., Sinhababu, D.P., Sahu, P.K., Satapathy, B.S., Shahid, M., Kumar, G.A.K., Jambhulkar, N.N., Nedunchezhiyan, M., Giri, S., Saurabh Kumar, Sangeeta, K., Nayak, A.K., and Pathak, H. (2019). *Rice-Fish Integrated Farming Systems for Eastern India*. NRRI Research Bulletin No. 17, ICAR-National Rice Research Institute, pp. 33+iii.

Ayyappan, S., Moza, U., Gopalkrishnan, A., Meenakumari, B., Jena, J.K., and Pandey, A.K. (2017). *Handbook of Fisheries and Aquaculture*. Directorate of Knowledge Management in Agriculture, Indian Council of Agricultural Research, pp. 275–301.

Behera, U.K., Jha, K.P., and Mahapatra, I.C. (2004). Integrated management of available resources of the small and marginal farmers for generation of income and employment in eastern India. *Crop Research*, 27, 83–89.

Bhanumurthy, V., Manjusree, P., and Rao, S. (2010). Flood Disaster Management, An eBook, Remote Sensing Applications, Chapter 12. In R.S. Roy, R.S. Dwivedi, and D. Vijayan (Eds.), *National Remote Sensing Centre*, ISRO, pp. 283–302, https://www.nrsc.gov.in.

Bhattacharyya, P., Sinhababu, D.P., Roy, K.S., Dash, P.K., Sahu, P.K., Dandapat, R., Nogi, S., Mohanty, S. (2013). Effect of fish species on methane and nitrous oxide emission in relation to soil C, N pools and enzymatic activities in rainfed shallow lowland rice-fish farming system. *Agriculture, Ecosystems and Environment*, 176, 53–62.

Das, A., Munda, G.C., Azad Thakur, N.S., Yadav, R.K., Ghosh, P.K., Ngachan, S.V., Bujarbaruah, K.M., Lal. B., Das, S.K., Mahapatra, M.K., Islam, M., and Dutta, K.K. (2014a). Rainwater harvesting and integrated development of agri-horti-livestock-cum-pisciculture in high altitudes for livelihood of tribal farmers. *Indian Journal of Agricultural Sciences*, 84(5), 643–649.

Das, A., Patel, D.P., Ramkrushna, G.I., Munda, G.C., Ngachan, S.V., Kumar, M., Buragohain, J., and Naropongla. (2014b). Crop diversification, crop and energy productivity under raised and sunken beds: Results from a seven-year study in a high rainfall organic production system. *Biological Agriculture & Horticulture*, 30(2), 73–87.

Datta, A., Nayak, D.R., Sinhababu, D.P., and Adhya, T.K. (2009). Methane and nitrous oxide emission from an integrated rainfed rice-fish farming system of eastern India agriculture. *Agriculture, Ecosystem and Environment*, 129(1–3), 228–237.

DES (2020). *Agricultural Statistics at a Glance 2019, Directorate of Economics and Statistics*. Department of Agriculture Cooperation and Farmers Welfare, Ministry of Agriculture Cooperation and Farmers Welfare, Government of India, pp. 40–92, 260–266.

Dugan, P., Dey, M.M., and Sugunan, V.V. (2006). Fisheries and water productivity in tropical river basins: Enhancing food security and livelihoods by managing water for fish. *Agricultural Water Management*, 80(1–3), 262–275.

FAO (1996). Report of the World Food Summit 1996, WFS 96/REP Part One, Food and Agriculture Organization, Rome, Italy, November 13–17, 1996. http://www.fao.org/3/w3548e/w3548e00. htm# doc09.

FAO, IFAD, UNICEF, WFP, and WHO. (2021). *The State of Food Security and Nutrition in the World 2021. Transforming Food Systems for Food Security, Improved Nutrition and Affordable Healthy Diets For All.* Food and Agriculture Organization, Rome, pp. 15–17. https://doi. org/10.4060/cb4474en.

FAO, WFP, and IFAD. (2012). *The State of Food Insecurity in the World 2012. Economic Growth is Necessary But Not Sufficient to Accelerate Reduction of Hunger and Malnutrition.* Food and Agriculture Organization, Rome, pp. 8–14.

FAOSTAT (2021). *FAO Statistics*. Food and Agriculture Organization, Rome. http://www.fao.org/statistics/en/.

Hai, V.M. and Van Dung, N. (2002). Litchi Production in Viet Nam. In Food and Agricultural Organization of the United Nations, Bangkok (Ed.), *Litchi Production in the Asia-Pacific Region*, Thailand, RAP Publication, pp. 114–119.

ICAR-RCER (2014). Sustainable Livelihood Improvement through Need-based Integrated Farming System Models in Disadvantaged Districts of Bihar. *Final Report*, NAIP-3 Subproject, ICAR-RCER, Patna, p. 94.

Khan, M.H., Bharti, P.K., and Kumar, S. (2014). Fish production through integration of livestock. *Indian Farming*, 64(1), 65–48.

Korikantimath, V.S. and Manjunath, B.L. (2008). Integrated farming systems for sustainability in agricultural production. *Proceedings of National Symposium on New Paradigms in Agronomic Research.* Indian Society of Agronomy, Navsari Agriculture University, pp. 279–281.

Kumar, S., Singh, S.S., Meena, M.K., Shivani, S., and Dey, A. (2012). Resource recycling and their management under integrated farming system for lowlands of Bihar. *Indian Journal of Agricultural Sciences,* 82(6), 504–510.

Lemma, A. (2013). Integrated poultry, fish and horticulture. In *Trends in the Conservation and Utilization of Aquatic Resources of the Ethiopian Rift Valley.* Paper presented at the 5th Annual Conference of the Ethiopian Fisheries and Aquatic Sciences Association (EFASA), Hawassa. Brook Lemma, Seyoum Mengistou, Elias Dadebo (EFASA), Zenebe Tadesse and Tadesse Fetahi, pp. 178–204.

Lemma, A. (2021). Integrated fish — poultry-horticulture — forage and fattening production system at Godino, Ada'a District, East Shoa Zone. *International Journal of Advanced Research in Biological Sciences,* 8(2), 15–25.

Mandal, K.G., Purbey, S.K., Singh, A.K., Bharti, P.K., Kumar, Ravi, Samal, S.K., and Banerjee, K. (2021). Perspectives on integrated farming in waterlogged ecosystems for ensuring food and nutrition security to farmers. *Indian Journal of Agronomy,* 66(5th IAC spl issue), S32–S43.

Mohanty, R.K. and Mishra, A. (2003). Rice-fish farming in the rainfed medium lands of eastern India. *Indian Farming,* 9, 10–13.

Mohanty, R.K., Verma, H.N., and Brahmanand, P.S. (2004). Performance evaluation of rice–fish integration system in rainfed medium land ecosystem. *Aquaculture,* 230(1–4), 125–135.

Nayak, P.K., Nayak, A.K., Tripathi, R., Panda, B.B., Kumar, A., Shahid, M., Kumar, U., Khannam, R., and Das, S.K. (2021). Rice-based integrated farming systems for doubling of farm income in Eastern India. *Indian Farming,* 71(4), 44–48.

Nedunchezhiyan, M., Sinhababu, D.P., Sahu, P.K., and Pandey, V. (2013). Growth and yield of sweet potato (*Ipomoea batatas* L.) in rice fallows: Effect of tillage and varieties. *Journal of Root Crops,* 39(2), 110–116.

Panda, M.M., Ghosh, B.C., and Sinhababu, P. (1987). Uptake of nutrients by rice under rice-cum-fish culture in intermediate deepwater situation (up to 50 cm water depth). *Plant & Soil,* 102, 131–132.

Patel, R.K., Srivastava, K., Pandey, S.D., Kumar, A., Purbey, S.K., and Nath, V. (2020). Productivity improvement of low-lying area with litchi *(Litchi chinensis)* based integrated system. *Indian Journal of Agricultural Sciences,* 90(4), 762–766.

Patra, B.C. and Sinhababu, D.P. 1995. Weed flora under rainfed lowland rice ecosystem with reference to rice-fish system. *Oryza*, 32(2), 121–124.

Preston, T.R. (2002). Towards local resources-based integrated crop-livestock systems. *LEISA Magazine*, April 2002, p. 26.

Saha, S. and Biswal, G.C. (2004). Rainfall distribution pattern and its implication for suitable crop planning under rainfed rice-based crop production system in Balasore district of Orissa. Paper presented in National Symposium on "Recent Advances in Rice-Based Farming Systems", held at CRRI, Cuttack during 17–19 November, pp. 80–81.

Saha, S., Singh, D.P., Sinhababu, D.P., Mahata, K.R., Behera, K.S. and Pandey, M.P. (2007). Rice-based production system for food and livelihood security in eastern coastal plain (lead paper). In International Symp. on "Management of Coastal Ecosystem: Technological Advancement and Livelihood Security" by ISCAR, October 27–30, 2007, Science City, pp. 3–4.

Sahoo, N., Roy Chowdhury, S., Brahmanand, P.S., Mohanty, Rajeeb K., Jena, S.K., Thakur, A.K., James, B.K., and Kumar, A. (2006). Integrated Management Approaches for Waterlogged Ecosystem. Research Bulletin No. 30. Water Technology Centre for Eastern Region (Indian Council of Agricultural Research), Chandrasekharpur, Bhubaneswar.

Singh, K., Bohra, J.S., Singh, Y., and Singh, J.P. (2006). Development of farming system models for the north-eastern plain zone of Uttar Pradesh. *Indian Farming*, 56(2), 5–11.

Sinha, R. (2008). Kosi: Rising waters, dynamic channels and human disasters. *Economic and Political Weekly*, 43(46), 42–46.

Sinha, R. (2009). Dynamics of the river system — The case of the Kosi River in north Bihar. e–*Journal Earth Science India*, 2(1), 33–45.

Sinha, R., Bapalu, G., Singh, L., and Rath, B. (2008). Flood risk analysis in the Kosi river basin, north Bihar using multi-parametric approach of Analytical Hierarchy Process (AHP). *Journal of the Indian Society of Remote Sensing*, 36(4), 335–349.

Sinhababu, D.P. and Majumdar, N. (1981). Evidence of feeding on brown planthopper, *Nilaparvata lugens* Stal. by common carp, *Cyprinus carpio* L. *Journal of Inland Fishery Society of India*, 13(2), 16–21.

Sinhababu, D.P. and Rajamani, S. (2000). Efficacy of insecticides and feasibility of their use in rainfed lowland rice-fish seed system. *Oryza*, 37(2), 129–132.

Sinhababu, D.P., Ghosh, B.C., Panda, M.M., and Reddy, B.B. (1983). Effect of fish on growth and yield of rice under rice-cum-fish culture. *Oryza*, 20(2), 144–150.

Sinhababu, D.P., Jha, K.P., Mathur, K.C., and Ayyappan, S. (1998). Integrated farming system: Rice-fish for lowland ecologies. In S.K. Mohanty *et al.* (Ed.), *Proceedings of the International Symposium on Rainfed Rice for Sustainable Food Security.* September 23–25, 1996. Central Rice Research Institute, p. 516.

Sinhababu, D.P., Panda, M.M., and Ghosh, B.C. (1992). Performance of rice and fish under the application of FYM and fish feed in rainfed intermediate lowland (0–50 cm). *Oryza*, 29, 221–224.

Sotto, R.C. (2002). Litchi production in the Philippines. In Food and Agricultural Organization of the United Nations, *Litchi Production in the Asia-Pacific Region*, RAP Publication, pp. 94–105.

Srivastava, A.P. (2018). Selected integrated farming system models for enhanced income. *Indian Farming*, 68(1), 13–16.

https://doi.org/10.1142/9789811296062_0002

Chapter 2

Out-of-Box Thinking and Skill-Building Exercises Encompassing Cornell Note-Taking Method to Facilitate Climate Smart Agriculture

Sudhanand Prasad Lal* and Akshay Singh

Post Graduate Department of Extension Education, College of Agriculture, Dr. Rajendra Prasad Central Agricultural University, Samastipur, Bihar, India
*sudhanand.lal@rpcau.ac.in

Abstract

Climate change is currently one of the most evident and concerning problems globally. The productivity of various crops is dropping at an alarming rate due to the rise in temperature resulting from climate change. Thinking outside the box is critical for solving this problem because it gives us a fresh perspective on innovatively solving existing problems. Training regarding climate change is conducted for farmers, trainers, etc., in which different note-taking methods are used, and note-taking is a typical part of training for retaining and better understanding the information given in the training. Different note-taking methods are suitable for different situations. For example, when comparing things with similar attributes, the charting method of note-taking is the best. Cornell note-taking is the most intuitive way of taking and recollecting notes for a long time. For addressing the problem of climate change out-of-the-box thinking and note-taking are quintessential, as these tools help with proper documentation, analysis, and reaching a feasible solution to the existing issues of climate change.

Keywords: Out-of-box thinking, Cornell note-taking, Climate change, Skill-building, SQ3R.

Introduction

Climate change is not a hoax, as evident from global warming, and falls in the most serious category. Global warming is now gravely threatening the yield of crops worldwide (Janni *et al.*, 2020). Unless CO_2 fertilization, successful adaptation, and genetic advancement are fully implemented, it is projected that every 1°C increase in the world mean temperature will result in a global yield reduction of 6.0%, 3.2%, 7.4%, and 3.1% for wheat, rice, maize, and soybeans, respectively (Zhao *et al.*, 2017). According to the Food and Agriculture Organization (FAO), the relative rates of production increase for the major cereal crops have declined. To maintain food security, crop output must rise with population growth (Bita and Gerats, 2013). It is predicted that by 2050, for the projected population of 9 billion food production would have to increase by 70%. To tackle these climate change challenges "out-of-box thinking" is crucial. In addition, "Note-Taking Method" is quintessential to record climate data properly. At the farm level, archives of climate data, including global historical weather, climate data, and station history data, may be kept if farmers clearly know how to record it. Moreover, Rainfall Time Series data analysis is based on historical data and consequently depends on proper note-taking and recording.

Out-of-Box Thinking

Out-of-the-box thinking is also called divergent thinking and creative thinking. *Merriam-Webster* defines out-of-the-box thinking as the exploration of creative and unconventional ideas that are not governed or confined by rules or tradition. It involves several factors and can lead to unique ideas and solutions associated with creativity (Lechelt, 2020). According to *Psychology Today*'s (Boyd, 2014) article titled "Thinking outside the box: A misguided idea," the concept was developed by psychologist J. P. Guilford, one of the earliest academic scholars to examine creativity in the early 1970s.

"When someone thinks outside the box, the box goes away," is a good way to summarize thinking outside the box. Outside-the-box thinking is not the same as lateral thinking. Out-of-the-box thinking is frequently compelled by leadership methods that make a living in the old box so unpleasant that deciding to leave is the only alternative. As a result, progress is in the hands of people who are willing and capable of making it. For many people, the problem with outside-the-box thinking is that it requires emotional understanding and management in addition to the more visible qualities of creativity, mental agility, and courage. Thinking beyond the box is tough as it means moving out of one's psychological comfort zone, opening to new viewpoints, and being willing to explore. It entails the removal of character armor as well as a personal "glasnost." Emotional management entails the ability to balance powerful lower-frequency emotions like fear and rage with more subtle high-frequency emotions like joy and happiness (Darn, 2006). Association of Psychological Science (2012) mentioned that thinking out of the box requires a free space and advocated that creativity is a highly sought-after skill.

Figures 1–4 depict typical examples of out-of-the-box thinking.

In Fig. 2, if one removes the matchstick in the right positive x-axis coordinate then a square will be formed at the center (Puzzle a Day,

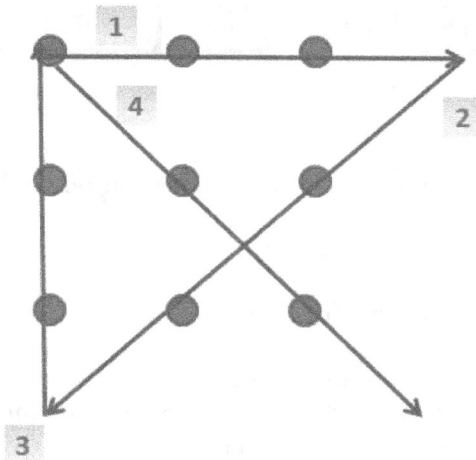

Fig. 1. Depicts 9 dots 4 lines puzzle.

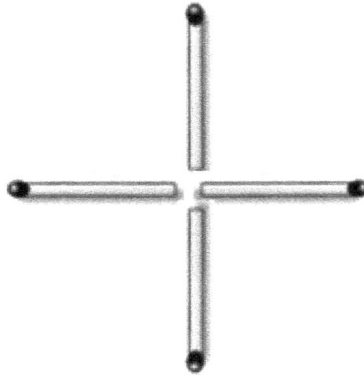

Fig. 2. Four matchsticks make a square.

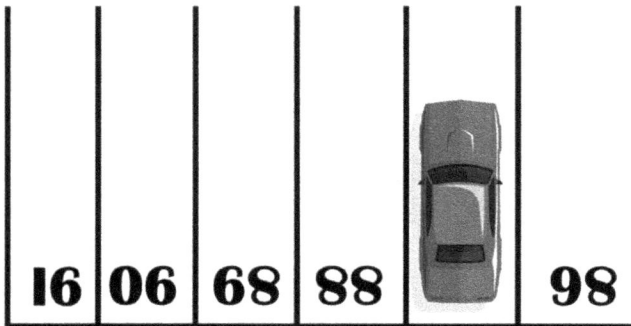

Fig. 3. What is the car's parking spot number?

2018). People usually think that a square can be made by reshuffling at least two matchsticks.

Cornell Note-Taking Method

If you want to succeed in classes, student groups, or other professions, you must be able to take good notes, and this is one of the most important skills (Clairehuchin, 2015; GoodNotes, 2022; Learning Hub, 2022). Taking notes is one thing, but taking effective notes is an art that will enable you to assimilate and integrate new

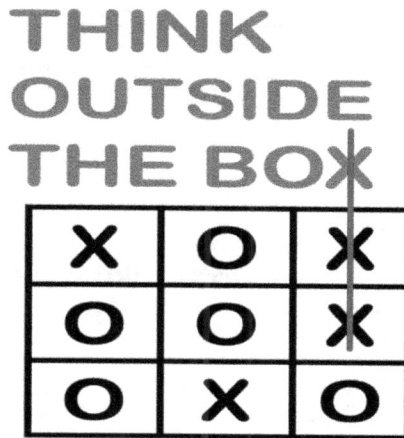

Fig. 4. Prominent example of out-of-the-box thinking.

information and analyze and organize the content you are learning. Cornell University named its note-taking method after Professor Walter Pauk, who invented it in the 1950s. The strategy is also effective in other situations. The Cornell Note-Taking System, or Cornell Notes, is another name for this commonly used procedure. In the Cornell method of taking notes, you can physically arrange your notes and assess your understanding of the subject matter without the need to create flashcards or any other kind of memory aid.

During the 1950s, a professor at Cornell University in Ithaca, New York, named Walter Pauk, developed a standardized style for taking notes in an organized manner prescribed as Cornell note-taking method. It features a page with four columns, and the topmost column contains a title and a specific date. After that, the page is divided into three main sections: two columns in the main part of the page that separate the body and margin. The margin on the left side (cue column) is where keywords, questions, and mnemonics from the content are written. One big column at the bottom is known as the footer, in which the summary is written. Note-taking area should be filled with main lecture notes, ideas, and key thoughts. Footer is used to write a concise summary of the information in the body and margin. This should be done at the end of every single page, not only at the conclusion of a lecture or class. Why is this important?

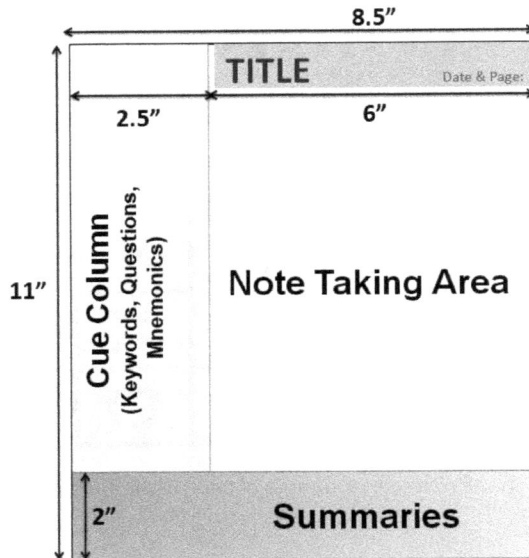

Fig. 5. Precise depiction of Cornell notes-taking method.

It should always be addressed in conclusions and _vice versa_ for the summary.

Mulder (2012) marked that the simplicity of the page layout is a strength of this strategy.

It begins at the top of the page by listing the name of the meeting, course, or seminar as well as the date and subject of the meeting, course, or seminar.

After that, the page is divided into three broad sections: the right-hand column is used for taking notes, the left-hand column is for jotting down questions, and the bottom section is for writing a summary (Fig. 5).

The Cornell Note-Taking Method and Abbreviations

Long sentences are discouraged in this strategy.

It's all about making short notes in the right-hand column using symbols, abbreviations, and acronyms that are simple to understand.

Making a list of acronyms and words before taking notes will lessen the hassle and make note-taking much simpler.

The right-hand column also contains important concepts, ideas, people, formulas, and graphs.

Short/brief and to-the-point notes.

On the left side of the page, questions are written using keywords related to the major points of the lecture or seminar.

The notes on the right may to be referred by numbers. A concise summary is written at the bottom of the page in your own words.

The bottom 5 cm of each page should be reserved for a summary or conclusion. Consolidating your understanding might be as simple as summarizing the notes in the bottom row.

Steps Involved in Taking Cornell Notes-Taking Method

Step 1: Complete the Cornell notes format and header.
Step 2: Arrange the notes on the right side.
Step 3: Reread and modify your notes.
Step 4: Make a list of key points to use as a starting point for questioning.
Step 5: Collaborate to exchange ideas (Stewart, 2022).
Step 6: Connect what you've learned to make a synthesized summary (Pauk & Owens, 2010; Quintus *et al.* 2012).

5 R's Format of Cornell Notes Note-Taking Method (Think Insights, 2022)

Record: Take notes using telegraphic sentences in the note-taking column during the lecture or meeting (Moviecultists, 2022; Purdue University, 2022). Keep track of as many important details and concepts from the lecture in the main column. Write legibly.

Questions: Soon after the session, construct questions based on the notes in the right-hand column. Writing questions aids in memory consolidation and helps to establish continuity, reveal relationships, and clarify meanings. It also creates an ideal environment for later analysis. Write a summary of these details and concepts in the Cue Column as soon as you can. Clarifying concepts through summarization improves continuity, retention, and understanding of relationships.

Recite: The best way to recite is by covering the note-taking column with a piece of paper or by palm. After that, speak aloud the information or concepts outlined in the question and cue column by simply focusing on them. Say over the facts and concepts of the lecture as completely as you can, using just your notes from the cue column and speaking in your own words rather than mechanically. Then, be sure what you said is true.

Reflect: Take ideas from your notes and use them as a foundation for your views on the course and its reflection on other courses. Reflection will aid in keeping ideas alive and from becoming inert or quickly forgotten. Think through the information and pose questions to yourself, for example:

What relevance do these facts have?
How do I put them into practice?
What guiding principle do they adhere to?
How do they fit into the situation I'm already familiar with?
What lies beyond them?

Review: Review your prior notes for a while. If you do, you'll be able to retain a lot for usage right away and discover any changes or best practices for work situations in the future. You will remember the majority of what you have studied if you spend 10 minutes each week quickly reviewing your notes (see Fig. 6).

Advantage

This results in more organized notes.

Using this technique, students can quickly and accurately identify key terms and ideas from lectures.

It also does away with editing and makes it easy to review by breaking important concepts down into questions.

It allows students to recall better all of the information they have acquired.

Assimilation of information happens quickly, leading to more efficient learning.

The lecture content can be thoroughly reviewed because of the clear overall picture.

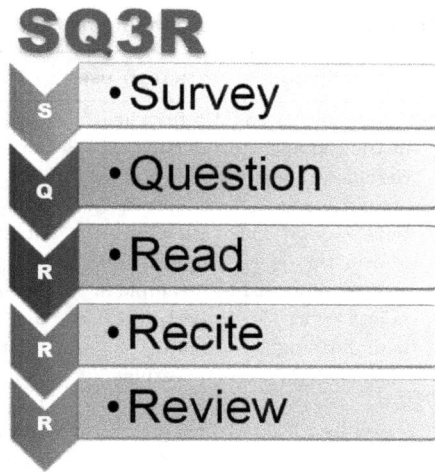

Fig. 6. SQ3R reading comprehension method.

Additionally, it enables them to apply the lecture information more quickly and get to the heart of the problem.

Disadvantage

For effective organizing, more thought is required in class (Studylib, 2022).

If the lecture is delivered too quickly, this approach will not work in precision.

Requires students to pay great attention to the lecture.

It is critical to examine notes on a daily basis.

It doesn't allow enough variety in a review session to improve learning and question applicability.

Table 1 shows a comparison between different note-taking methods.

Conclusion

The impact of climate change all around the globe is quite evident. In the agricultural world, paddy output dropped by 6.2% when the average temperature increased by 1°C during the rice growing season. A decrease of 7.1–8.0% was observed in total milled rice yield,

Table 1. Different note-taking methods.

S. No.	Method	Description	When to use	Pros	Cons
1.	Outline	The title is written at the top and the main points are written in bullets; additional information is written in nested points under the main heading.	Useful when jotting down information quickly, for example in lectures and meetings.	Most structured and visually organized. Highlights key points and groups related points together.	Not suitable for subjects that have many charts, visuals, and diagrams.
2.	Cornell	Created by Prof. Walter Pauk at Cornell University in the 1950s. Three main sections: • Key thoughts are written in the notes section. • Questions related are written in the Que section. • After the lecture, concise summary is written in the Summary section.	This method is very effective for studying because it makes revising so easy.	Writing the summary enhances your knowledge of the topic. Helps in extracting main ideas, and it is logically organized.	More effort is required while taking notes. Time is required to set up the page.
3.	Boxing	It is a highly visual note-taking method. It gives you an at-a-glance summary of a topic. Each section or sub-topic resides in its labeled box.	Highly recommended during the time of revision.	Gives you a summary at a glance.	Not suitable for lectures where notes need to be written quickly. Drawing boxes free hand can be a hassle.

Table 1. (*Continued*)

S. No.	Method	Description	When to use	Pros	Cons
4.	Charting	This is a fantastic technique to group various things or ideas that have several features.	When comparing objects based on a certain set of attributes, charting is helpful.	Series of items are summarized systematically, which is great for comparison.	Not suitable for taking notes that follow a plot or development of material or notes that are more linear.
5.	Mapping (Mind map)	This method allows you to organize notes by dividing them into branches and establishing the relation between topics. It visually looks like neurons of the brain.	When a lot of explanation is needed for a single point, this approach works perfectly. It also functions when your notes are organized into a tale or a linear progression.	Easy to create. Easily establishes connections between the information.	If you have several informational branches, you can run out of space on the page.

a decrease of 9.0–13.8% was observed in head rice yield, and a decrease of 8.1–11.0% was observed in total milling revenue. Therefore, to tackle the challenge of climate change, yield, and revenue loss, "Out-of-Box thinking" is a must. In addition, "Note-Taking Method" is quintessential to record climate data properly and precisely.

References

Association for Psychological Science. (2012, January 24). To "think outside the box," think outside the box. *ScienceDaily*. Retrieved on February

25, 2022 from www.sciencedaily.com/releases/2012/01/120123175800. htm.

Bita, C.E. and Gerats, T. (2013). Plant tolerance to high temperature in a changing environment: Scientific fundamentals and production of heat stress-tolerant crops. *Frontiers in Plant Science*, 4, 273. doi: 10.3389/fpls.2013.00273

Boyd, D. (2014). Thinking outside the box: A misguided idea. *Psychology Today*. Retrieved on February 25, 2021 from https://www.psycholog ytoday.com/us/blog/inside-the-box/201402/thinking-outside-the-box-misguided-idea and https://drewboyd.com/.

Clairehuchin. (2015). A little story of note-taking. *CultureXchange*. https://culturexchange1.wordpress.com/2015/05/29/a-little-story-of-note-taking/.

Darn, S. (2006). Thinking outside the teacher's box. *Semantic Scholar*. Retrieved on February 24, 2021 from https://www.semanticscholar. org/paper/Thinking-outside-the-Teacher's-Box.-Darn/7658e6895ffca 10b42040cbb2c9872a52b07ac96#citing-papers.

GoodNotes. (2022, September 19). The best note-taking methods for college students & serious note-takers. https://medium.goodnotes.com/the-be st-note-taking-methods-for-college-students-451f412e264e.

Janni, M., Gullì, M., Maestri, E., Marmiroli, M., Valliyodan, B., Nguyen, H.T., Marmiroli, N., and Foyer, C. (2020). Molecular and genetic bases of heat stress responses in crop plants and breeding for increased resilience and productivity. *Journal of Experimental Botany*, 71(13), 3780–3802. https://doi.org/10.1093/jxb/eraa034.

Learning Hub. (2022). Note taking methods. The University of Auckland. https://www.learninghub.ac.nz/study-skills/readings/note-taking/.

Lechelt, A. (2020). Research paper: Thinking out of the box. *International Coach Academy*. Retrieved on February 25, 2021 from https://coachcampus.com/coach-portfolios/research-papers/am y-lechelt-thinking-outside-the-box/.

Mulder, P. (2012). Cornell note taking method. Toolshero. Retrieved on February 25, 2022 from https://www.toolshero.com/personal-develop ment/cornell-note-taking-method/.

Pauk, W. and Owens, R.J.Q. (2010). How to study in college. *The Cornell System: Take Effective Notes*, 10th edn. Boston, MA: Wadsworth, pp. 235–227, Chapter 10. Retrieved on February 25, 2022 from https://lsc .cornell.edu/how-to-study/taking-notes/cornell-note-taking-system/.

Puzzle a Day. (2018). Move one matchstick to make a square. Retrieved on February 25, 2022 from https://puzzleaday.wordpress.com/2018/07/0 9/move-one-matchstick-to-make-a-square/.

Quintus, L., Borr, M., Duffield, S., Napoleon, L., and Welch, A. (2012). The impact of the Cornell note-taking method on students' performance in a high school family and consumer sciences class. *Journal of Family and Consumer Sciences Education*, 30(1), 27–38.

The University of Tennessee Chattanooga. (2022). Common note-taking methods. https://www.utc.edu/enrollment-management-and-student-affairs/center-for-academic-support-and-advisement/tips-for-academic-success/note-taking.

Think Insights (2022). Cornell method: A method to take great notes. Retrieved on February 25, 2022 from https://thinkinsights.net/consulting/cornell-method-great-notes/.

UMFK (2022). Cornell note taking method. University of Maine at Fort Kent. https://www.umfk.edu/student-success/academic-support/notes/.

Williams, E. (2022). Cornell note taking system: What is it and how to use it? *Wondershare*. https://pdf.wondershare.com/mobile-app/cornell-note-taking-system.html.

Zhao, C., Liu, B., Piao, S., Wang, X., Lobell, D.B., Huang, Y., Huang, M., Yao, Y., Bassu, S., Ciais, P., Durand, J.L., Elliott, J., Ewert, F., Janssens, I.A., Li, T., Lin, E., Liu, Q., Martre, P., Müller, C., Peng, S., Peñuelas, J., Ruane, A.C., Wallach, D., Wang, T., Wu, D., Liu, Z., Zhu, Y., Zhu, Z., and Asseng, S. (2017). Temperature increase reduces global yields of major crops in four independent estimates. *Proceedings of the National Academy of Sciences of the United States of America*, 114, 9326–9331.

Webliography

Moviecultists. (2022). Why use Cornell notes? Retrieved on February 25, 2022 from https://moviecultists.com/why-use-cornell-notes.

Purdue University. (2022). Note-taking tips & methods to improve your notes. Retrieved on February 25, 2022 from http://www.sic.edu/files/uploads/group/34/PDF/ASC_Handouts_ImprovingNotetaking.pdf.

Stewart, L. (2022). 7 steps for successful note taking Cornell method. *Become a Writer Today*. Retrieved on February 25, 2022 from https://becomeawritertoday.com/note-taking-cornell-method/.

Studylib. (2022). Method description advantages disadvantages example Cornell. Retrieved on February 25, 2022 from https://studylib.net/doc/18161694/method-description-advantages-disadvantages-example-cornell.

Chapter 3

Artificial Intelligence for Climate Smart Agriculture

M.S. Kulshrestha* and Suvarna Dhabale

*Department of Basic Science and Humanities,
B.A. College of Agriculture, Anand Agricultural University,
Anand, Gujarat, India*
*kush122003@yahoo.co.in

Abstract

Climate-smart agriculture (CSA) is crucial for addressing the challenges posed by climate change in the agricultural sector. As climate variability intensifies, there is an urgent need for innovative solutions to enhance agricultural resilience, productivity, and sustainability. Artificial Intelligence (AI) emerges as a powerful tool in this context, offering transformative capabilities to revolutionize agricultural practices. The integration of artificial intelligence (AI) into climate-smart agriculture presents a paradigm shift in the way we approach farming. AI technologies, including machine learning, remote sensing, and data analytics, enable real-time monitoring and prediction of climatic patterns, pest infestations, and crop health. By harnessing the power of big data, AI assists farmers in making informed decisions related to crop selection, irrigation, and resource management. Moreover, AI-driven precision agriculture optimizes resource usage by precisely tailoring inputs such as water, fertilizers, and pesticides, minimizing waste, and environmental impact. Collaborative AI platforms facilitate knowledge sharing among farmers, fostering a community-driven approach to climate resilience. It highlights the potential of AI to revolutionize climate-smart agriculture, offering a sustainable path forward for farmers to adapt to changing environmental conditions while ensuring global food security. The integration

37

of AI into agriculture not only enhances productivity but also contributes to the broader goal of creating a resilient and sustainable food system in the face of climate challenges.

Keywords: AI, Climate change, CSA.

Climate Smart Agriculture

Climate-smart agriculture (CSA) is often referred to as climate-resilient agriculture (CRA). This is an integrated method of landscape management that tries to help the adaptation of agricultural management practices, animals, and crops to the effects of climate change. Also, where feasible, to reduce those effects by keeping low greenhouse gas emissions from agriculture taking into account the growing world population for assurance of food security (FAO, 2013). Therefore, in addition to carbon farming and sustainable agriculture enhancing agricultural productivity is crucial. The Food and Agriculture Organization (FAO) aims to increase agriculture, forestry, and fisheries production and sustainability, and CSA is in line with FAO's goal for sustainable food and agriculture (FAO, 2019; DownToEarth, 2021).

The three pillars of CSA are increasing agricultural output and incomes, adapting to climate change and enhancing resilience, and lowering or eliminating agricultural greenhouse gas emissions. CSA provides a number of solutions for plant and crop-related challenges.

Strategies and Technologies for Climate Change Adaptation

Climate change adaptation entails making the required alterations and changes to reduce the negative repercussions of climate change (or capitalizing on the good aspects). The following adaptation strategies are outlined in agricultural and livestock adaptation methodologies (DownToEarth, 2021).

Tolerant crops: Climate-resilient crops and agricultural types have subsequently been developed to address these challenges. Greengram (BM 2002-1), chickpea (BDN-708), and pigeon pea (BDN-708) early-ripening and drought-tolerant cultivars were planted in a few farmer's fields in Aurangabad, Maharashtra, to achieve inadequate downpour circumstances (rainfall of 645 mm).

Tolerant breeds in livestock and poultry: The animals have the ability to withstand the climate change effect. Breeds that are local or indigenous have the mindset that they can feed themselves. Animals in nomadic systems provide their owners cues on when to relocate in search of fresh meadows. Indigenous breeds across the world have distinctive traits that have been adapted to extremely particular ecosystems. Drought resilience, thermoregulation, long-distance walking abilities, maternal instincts, the capacity to consume and digest poor feed, and disease resistance are some of these animals' distinctive traits. These varieties of cattle may not be particularly productive in terms of meat or milk production but they are highly resilient to the unpredictability of the environment and have small environmental impacts.

Feed management: Improving feeding systems as a response strategy might inadvertently boost the productivity of livestock farming. Changing the amount or frequency of feedings, changing the diet's composition, feeding agroforestry species to animals, and teaching farmers how to produce and store feed for different agroecological zones are some of the feeding techniques.

Water management: Farmers may help lessen the impact of climatic change by using water-smart technologies such as raised beds watered by furrows, micro-irrigation, rainwater collection structures, cover crops, greenhouses, laser land leveling, reusing wastewater, deficit irrigation, and drainage management. In order to create and develop affordable and ecologically friendly water-conserving technologies, several international organizations, national governments' research institutes, farmers' groups, non-profits, and commercial entities have been concentrating their efforts globally.

Agro-advisory: Agro-advisory services are farm decisions made in reaction to historical, present, and future weather changes. It includes agronomy, pest and disease control, water management, and input management. Weather-sensitive crops, their weather-sensitive phases, and weather-sensitive farm activities are the fundamental elements for preparing weather-based agro-advisories. It is an integrative strategy. Tamil Nadu and many other states have already adopted responsive farming, which has reduced risk and increased output.

Organic carbon in soil: Soil organic carbon (the carbon present in organic matter in soil) is critical for soil health, fertility, and ecosystem services such as food production, hence preserving and restoring it is critical for long-term development. Carbon-rich soils are more likely to be productive and capable of filtering and purifying water. A major factor in climate change is soil organic carbon.

National Programs for Climate Change Adaptation

The Government of India has undertaken the convergence of several policy programs and sectoral strategies to achieve synergy and effective utilization of available resources. One of the eight missions within the National Action Plan on Climate Change (NAPCC), the National Mission of Sustainable Agriculture was adopted in 2010 to encourage the wise management of resources.

Pradhan Mantri Krish Sinchai Yojana (PMKSY) was introduced in 2015 to solve the water source problem and provide a long-term answer to increase yields per drop. It encourages micro/drip irrigation to save the maximum water possible. The Paramparagat Krishi Vikas Yojana mission was carried out in collaboration with the Indian Council of Agricultural Research and state governments of India to fully adapt climate-smart practices and technology. The Government of India (GOI) established the Green India Mission in 2014 under the auspices of NAPCC with the primary goal of conserving, improving, and restoring India's dwindling forest covers, thereby decreasing the harmful consequences of climate change. The Soil Health Card program was established by the GOI with the goal of analyzing soil samples from clusters and advising farmers on the fertility of their land. Neem-Coated Urea was further introduced to reduce the overuse of urea fertilizers, preserving soil health and feeding plants with nitrogen.

Artificial Intelligence

Software innovations that enable a robot or computer to behave and think like a person are referred to as Artificial Intelligence or AI. Computer systems that can carry out activities that ordinarily require human intellect have been developed using the theory of artificial intelligence. It is known as machine learning. The phrase "artificial intelligence" was first used by American computer scientist and cognitive scientist John McCarthy (1927–2011). Machine learning (ML) is a subgroup of AI. That is, all machine learning considers AI but not all AI does the same. For example, symbolic logic — rules engines, expert systems, and knowledge graphs — may all be classified as AI, but none of these are machine learning.

Artificial neural networks (ANNs) are used in ML, a branch of AI, to simulate human decision-making. Computers can learn from big datasets autonomously and without programming thanks to ML. Deep learning (DL), one of the several machine learning strategies, goes one layer deeper. Deep neural networks are used in deep learning to extract patterns from enormous volumes of data.

Scope of AI in agriculture

AI and ML are being rapidly used in agriculture, both in terms of agricultural goods and in-field farming practices. Cognitive computing, in particular, is poised to become the most disruptive technology in agricultural services because of its ability to comprehend, learn, and adapt to various situations (based on learning) to boost efficiency. The use of AI technologies in the agricultural sector is known as precision agriculture or satellite farming.

Yield prediction of paddy using multiple linear regression, principal components analysis, and ANN models are seen in Anand, Gujarat.

Data used

Kharif rice yield data (kg/ha) from 1988 to 2016 were obtained from the Directorate of Economics and Statistics, Government of Gujarat, Gandhinagar in accordance with the study's particular goals. Maximum temperature (X1), minimum temperature (X2), relative humidity (X3), wind speed (X4), and total rainfall (X5) were used as inputs to investigate the impact on *kharif* rice crop yield as output. Weekly meteorological data from the first fortnight before sowing to the end of the reproductive stages were used to build statistical models for the *Kharif* season. Weather data was used from May–October (22–41st standard meteorological week (SMW).

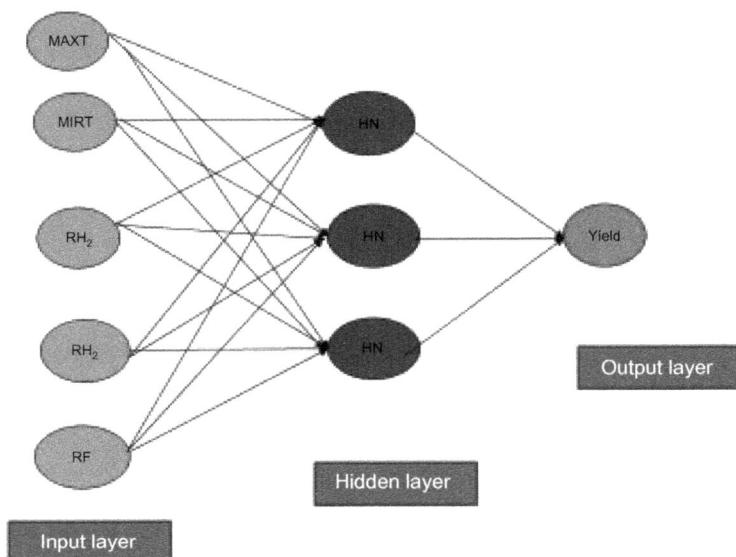

Fig. 1. ANN with feed-forward and backward propagation layer and connection diagram.

In this model, weekly weather variables of 17 weeks have been used. Weather indices (weighted and unweighted along with their relationships with each other) were also introduced. Multiple Linear Regression (MLR) model was designed with 30 weather indices (15 unweighted and 15 weighted indices) as a standalone variable and paddy yield as dependent variables. Whereas 30 weather indices (unweighted and weighted, J = 0, 1) have been used to build the principal components analysis (PCA). ANN consists of input layer, hidden layer, and output layer. Input and output layers have one neuron (Fig. 1).

Result and discussion

The major goal was to determine the strength of the relationship between rice yield and weekly meteorological factors. During the rice cropping season (from July 2 to October 14), the other week-wise correlation coefficient between yield and meteorological factors was determined to be non-significant. The value of "r" ranges from 0.45 to 0.46, suggesting that each character does not account for more than 46% of the variation in yield. The results were statistically analyzed using four statistical procedures like MPAE (Mean Absolute Percent Error), MBE (mean bias error), RMSE (Root Mean Square Error), and r (correlation coefficient) as given in Willmott (1981). Statistical model validation was done between MLR, PCA-MLR, and ANN. Among the comparisons chosen, the best fit models were chosen based on r, MPAE, MBE, and RMSE. The details of the comparative study were given. The value of r for MLR, PCA-MLR, and ANN are 0.5, 0.5, and 0.7 respectively, and a value of RMSE 168.9, 229.3, and 131.4, respectively. The MPAE percent varies from 4.4 to 8.6. However, Table 3 showed model PCA-MLR which was chosen as the best fit model based on the highest r and lower RMSE. Its MPAE was less than 10% and highest among other models. Therefore, this model result shows good agreement between actual and predicted yield. It can be concluded PCA-MLR as a best fit model for the Anand district as compared to other models (see Tables 1–3).

Table 1. Actual and predicted paddy yield (kg/ha) by MLR.

Year	Actual	Predicated	RMSE	r	Mean absolute percent error (MAPE) (%)	Mean bias error (MBE)
2013	2324	2198.5				
2014	2560	2486.6	168.994	0.46	5.9	−92.75
2015	2485	2596.8				
2016	2585	2301.1				

Table 2. Actual and predicted paddy yield (kg/ha) by ANN.

Year	Actual	Predicated	RMSE	r	Mean absolute percent error (MAPE) (%)	Mean bias error (MBE)
2013	2324	2334.3				
2014	2560	2366.3	131.41	0.7	4.4	−106.2
2015	2485	2398.3				
2016	2585	2430.3				

Table 3. Actual and predicted paddy yield (kg/ha) by PCA-MLR.

Year	Actual yield	Predicated yield	RMSE	r	Mean absolute percent error (MAPE) (%)	Mean bias error (MBE)
2013	2324	2100.3				
2014	2560	2238.1	229.345	0.471	8.6	−167.4
2015	2485	2579.6				
2016	2585	2366.4				

Conclusion

Estimations of rice crop production for the Anand district were effectively computed using forecast techniques like MLR, PCA-MLR, and PCA five weeks before the actual harvest. The low error percentage and strong correlation between observed and simulated rice production were evidence of the efforts. Validation of these models was done for years 2014–2015, 2015–2016, and 2016–2017. It can be concluded that among different statistical models for the Anand district, the PCA-MLR approach was chosen as the best fit model based on the greatest r and lowest RMSE.

Monthly Rainfall Predictions by Double Fourier Series and Artificial Neural Networks

Data used

An attempt has been made here to predict the summer monsoon monthly rainfall by two nonlinear models namely, Double Fourier Series (DFS) and ANN of Anand station of Gujarat, India. DFS was used with two inputs monthly, i.e., maximum temperature and relative humidity and one output that is $z = f(x,y)$. In this process, first different coefficients of DFS were determined by using historical data series of 56 years from 1958 to 2013 and 5 years of monthly rainfalls from 2014 to 2018 predicted. These values were validated with actual occurred rainfalls. Here, ANN consist of input layer, hidden layer, and output layer (Fig. 2). Input layer and output layer have

Table 7. Prediction of September from 2014 to 2018 by DFS model.

Year	Actual rainfall (mm)	Predicted rainfall (mm)	Difference between (A-P)	Mean absolute percentage error (%)	Root mean square error (RMSE)
2014	348.2	440.2	−92		
2015	79.1	46.45	32.65		
2016	178.4	173.69	4.71	37.2%	52.4
2017	97.8	89.52	8.28		
2018	60.0	124.17	−64.17		

Table 8. Prediction of June from 2014 to 2018 by ANN model.

Year	Actual rainfall (mm)	Predicted rainfall (mm)	Difference between (A-P)	Mean absolute percentage error (%)	Root mean square error (RMSE)
2014	4.8	0	4.8		
2015	92.0	96.0	−4.0		
2016	41.8	29.4	12.4	31.5%	10.1
2017	125.4	113.6	11.8		
2018	95.0	81.6	13.4		

Table 9. Prediction of July from 2014 to 2018 by ANN model.

Year	Actual rainfall (mm)	Predicted rainfall (mm)	Difference between (A-P)	Mean absolute percentage error (%)	Root mean square error (RMSE)
2014	438.0	431.0	7.0		
2015	309.0	300.0	9.0		
2016	80.0	79.0	1.0	2.1%	8.2
2017	351.0	359.0	−8.0		
2018	450.0	462.0	−12.0		

both the models were good. However, model ANN is more favorable for predicting rainfall. Predicted rainfall from June to September for 2014 to 2018 were listed in Tables 4–9 with their difference from actual occurred rainfall, respectively. September 2014 received rainfall of 348.2 mm which was nearly three times higher than normal

Table 10. Prediction of August from 2014 to 2018 by ANN model.

Year	Actual rainfall (mm)	Predicted rainfall (mm)	Difference between (A-P)	Mean absolute percentage error (%)	Root mean square error (RMSE)
2014	175.6	187.0	−11.4		
2015	309.0	300.0	9.0		
2016	231.4	223.0	8.4	4.7%	9.7
2017	142.4	155.0	−12.6		
2018	332.2	338.0	−5.8		

Table 11. Prediction of September from 2014 to 2018 by ANN model.

Year	Actual rainfall (mm)	Predicted rainfall (mm)	Difference between (A-P) (mm)	Mean absolute percentage error (%)	Root mean square error (RMSE)
2014	348.2	300	48.2		
2015	79.0	84.4	−5.4		
2016	178.4	180	−1.6	7.6%	22.1
2017	97.8	100	−2.2		
2018	60.0	51.36	8.64		

of the considered data series and therefore the model was unable to predict it.

Conclusion

Predicted monthly rainfall from June to September for 2014 and 2018 have greater accuracy and obtained MAPE was less than 10% except for June. For the prediction of monthly rainfall from June to September both the models were good but ANN is more preferable except for the month of July.

References

DownToEarth (2021). https://www.downtoearth.org.in/blog/agriculture/climate-resilient-agriculture-systems-the-way-ahead-75385.

Food and Agriculture Organization of the United Nations. (2013). *Climate-Smart Agriculture Sourcebook*. FAO.

Food and Agriculture Organization of the United Nations. (2019). *Climate-Smart Agriculture*. FAO.

Willmott, C.J. (1981). On the validation of models. *Physical Geography*, 2, 184.

Chapter 4

Extension Strategies for Climate Resilient Agriculture

R.N. Padaria* and Sudhanand Prasad Lal[†]

*Division of Agricultural Extension ICAR-IARI,
New Delhi, India

[†]Post Graduate Department of Extension Education,
Dr. Rajendra Prasad Central Agricultural University,
Pusa, Samastipur, Bihar, India
*rabi64@gmail.com

Abstract

Throughout the world, climate change has become a crucial issue while forming national policies as it poses a serious threat to food security in many countries. Even the United Nations has recognized climate change as a serious concern and included it in one of the 17 Sustainable Development Goals (SDGs). Nowadays, agriculture is more knowledge-intensive than ever before. However, farmers still suffer significant losses due to climatic risks because of their inability to access real-time, need-based, and accurate information. The key to successful adaptation to climate risks and vulnerabilities is the implementation of early warning systems and the rapid dissemination of contingency technology options. In recent years, mobile-based advisory services have shown great promise in prompt dissemination of information. Concerted efforts are being made toward the dissemination of technologies for effective adaptation such as climate-smart varieties besides being able to resist drought, floods, salinity, heat, disease, and pest infestation. The plants can also utilize water and nitrogen efficiently, change land management practices, adjust planting dates in order to minimize the effects of temperature increase and avoid heat stress during flowering periods. With the introduction of appropriate wheat varieties, *viz.*, HI 1500, HI 1531, HI 1544, and HI 8627 developed by IARI Regional Station, Indore, the farmers could secure an 8–20% higher yield compared to local check. In addition, the adoption of Extension Strategies suggested by pioneering climate change adaptation projects, i.e., National Innovations in Climate Resilient Agriculture (NICRA), will give further impetus to Climate Resilient Agriculture.

Keywords: Climate resilient agriculture, Extension strategies, Global Hunger Index, NICRA, Sustainable Development Goals.

Introduction

Climate change has become one of the most important agendas of policy formulation across the globe as it could pose a serious threat to food security. As the population is growing faster and the natural resources are sharply shrinking, there would be limited ways and means to address the increasing climatic risks. Poor adaptation capacity and high exposure to extreme weather events make the Indian agriculture system highly vulnerable to climate change. The small and marginal farmers operating at a subsistence scale would face serious threats to productivity, food security, and income security. Although India has the capacity to cover 2.347 times the moon distance if one Jute Sack of stored food grains in the warehouse is staked one over another (Lal *et al.*, 2022), according to the 2021 Global Hunger Index, India ranked 101st out of the 116 countries. Therefore, there is a growing significance of addressing the adverse impact of climate change as it has become one of the most challenging hurdles in attainment of the developmental goals. Even, the United Nations has recognized climate change as a serious concern and included it in one of the 17 Sustainable Development Goals (SDGs). Taking urgent action to combat climate change is one of the most important issues that need to be addressed. A number of specific targets of SDG 13 include the improvement of resilience and adaptive capacity in all countries compared to climate-related hazards and natural disasters (United Nations Publication, 2019). Aside from integrating climate change measures into national policies, strategies, and planning, climate change mitigation, adaptation, impact reduction, and early warning mechanisms should be promoted, as well as implementing measures to mitigate climate change, adapt, reduce impact, and enhance awareness.

Depending upon the nature of climatic stresses and availability of resource endowments, farmers have been taking initiatives for autonomous adaptation measures, which may be at times insufficient. Planned adaptation strategies in agriculture too are being promoted. Concerted efforts are being made toward the dissemination of technologies for effective adaptation such as climate-smart varieties. Besides being able to resist drought, floods, salinity, heat, disease, and pest infestation, these plants also possess the ability to utilize water and nitrogen efficiently, change land management practices, adjust planting dates in order to minimize the effects of temperature increases and avoid heat stress during flowering periods, and use resource-conserving technologies (Padaria *et al.*, 2017a) to reduce the effects of temperature increase, rainwater harvesting and efficient on-farm water management, crop diversification, brown manuring, site-specific nutrient management, Integrated Pest Management, and improved weather forecasting and crop insurance schemes.

Understanding Social and Behavioral Dimensions of Climate Change Adaptation

Despite the availability of effective technological options, mainstreaming adaptation remains a matter of concern. Socioeconomic factors as well as climatic factors both affect the decisions of adaptations to climate change. Behavioral orientation and different capitals like natural, physical, financial, human and social, land tenancy, and collectivization influence individual decisions for climate change adaptation.

A survey of farmers' perceptions revealed that a majority perceived climate change as a reality (Padaria *et al.*, 2017b). As far as the perception of farmers regarding adaptation is concerned, it is a positive and encouraging reflection. It is anticipated that they will take corrective measures to minimize the risks. A metaphysical interpretation of climate change, such as the curse of God, can hamper the acceptance of adaptation plans among farmers. Likewise, a pessimistic mindset, the notion that it is a temporary phenomenon, and nature's revenge can also impede the implementation of adaptation plans.

Human capital is of immense importance. Knowledge about climate change and technological options is vital for adaptation. Skills primarily affect the efficiency and efficacy of individuals and organizations for increased adoption and mitigation efforts of climate change. Improvement in skill would certainly improve the level of adaptation besides creating human wealth and resource conservation. Social capital is vital for collective action toward adaptation. It acts as the agent to influence the decision-making of people and their social connectedness for climate change adaptation. It also discovers how relatives, neighbors, civil society organizations, businesses, or government agencies affect climate change adaptation. If the social capital is stronger to influence the people in decision-making there would be fewer possible barriers for climate change adaptation and hence social capital plays one of the important roles in climate change adaptation.

Therefore, technology transfer endeavors should also engage in the promotion of human as well as social capital among the farmers. Social capital not only activates effective management of natural and physical capital but also increases the knowledge and skills of an individual through social learning. Social learning in agriculture refers to learning about different aspects of agriculture through social interaction between farmers and individuals in their network of contacts (Antwi, 2015). An agricultural farmer may benefit from such learning at various stages of his or her agricultural career, including when choosing crops and inputs; when applying inputs; or whenever he or she adopts a new technology. Thus social learning has an important role in adaptation. Capacity building and cooperation can be achieved if local communities diagnose and define problems or risks and then devise and disseminate solutions. Participatory processing and sharing of experiences facilitate social ratification as well as internalization of the knowledge.

It is imperative that vulnerable communities must work together in order to achieve collective adaptation that will result in effective individual adaptation decisions and to facilitate the sharing of risks between the villages and clusters of villages (CESCRA, 2014). Availability of quality seeds, planting materials, and fodder is often a serious constraint during climatic stress, either drought or flood. Promotion of community seed and fodder banks could resolve these issues at the local level.

With collective action, introduction of interventions, *viz.*, development or/and rejuvenation of small-scale water resources, i.e., water harvesting system (WHS), pond renovation, and deepening of open wells, becomes easier. *Jal Sahelis* or women water warriors have shown the way to withstand the pressures of drought through effective water management. *Zoba* system, a traditional system of Nagaland, has been effective in water harvesting and practicing settled agriculture while eschewing shifting cultivation.

Gender is another critical dimension for climate change adaptation. The growing feminization of agriculture makes women even more vulnerable to climatic risks. Lack of access to land, capital, training, and other extension services impedes adaptation among women farmers. Gender-sensitive extension approach and gender mainstreaming with committed gender budgeting in agricultural development are critical for effective adaptation.

Migration has become one of the major measures for adaptation to climate change. Though the livelihood security is attained through remittances partly, the household development is retarded. Agri-based as well as non-farm-based enterprises need to be promoted for reducing migration from villages.

Transferring Appropriate Technologies

A strong linkage between research and extension could facilitate better screening of appropriate technologies for adaptation. The tribal farmers in district Dhar of Madhya Pradesh often suffer from risks of drought and high temperatures in wheat cultivation. However, with the introduction of appropriate wheat varieties, *viz.*, HI 1500, HI 1531, HI 1544, and HI 8627 developed (Naresh *et al.*, 2014) by IARI Regional Station, Indore, the farmers could secure 8–20% higher yield in comparison to local check.

Though commendable efforts have been made by research organizations and state departments of agriculture, and district levels, contingency plans are available. During the drought of 2012, a Drought Contingency Plan was developed by the IARI team for the Mewat district in Haryana to deal with the continuing dry conditions that were predicted (CESCRA, 2014). Since due to drought the pearl millet crop could not be taken up, the farmers were suggested to adopt the contingency plan of Fallow-Early Mustard-Wheat. With the sowing of early mustard in the last week of September and harvesting in the last week of December, while taking late sown wheat (Var. WR 544), the farmers could not only offset the loss of *kharif* crop due to drought but also secure higher income than the traditional cropping pattern. However, it is prudent to develop location-specific contingency plans as per the micro-situations and devise mechanisms for their application by the farmers during the distress of climatic risks. For accelerated adoption of any technology provided by the institutes, care should be taken to develop such technologies that are socially, economically, and technically feasible for the farmers.

Real-time advisory is critical to optimal as well as contingency planning. Nowadays, agriculture is more knowledge-intensive than ever before. However, farmers suffer significant losses due to climatic risks because of their inability to access real-time, need-based, and accurate information. The key to successful adaptation to climate risks and vulnerabilities is the implementation of early warning systems and the rapid dissemination of contingency technology options. In recent years, mobile-based advisory services have shown great promise in a prompt dissemination of information (CESCRA, 2014). Pusa mKRISHI, a mobile-based advisory service, developed with a public–private partnership approach of IARI and TCS, Mumbai, has proven to be effective in providing weather alerts and weather-based customized and appropriate advisory through two-way information-seeking and provisioning system. It led to effective and informed decision-making by farmers in the experimental areas of Mewat (Haryana), Dhar (Madhya Pradesh), and Ganjam (Odisha).

The e-learning system has emerged as an effective means of acquisition of knowledge. Extension approaches have begun to utilize e-learning for effective dissemination of technical know-how and do-how. However, there is a dire need to enhance the digital competencies of farmers as well as change agencies. Grassroot innovations too hold immense potential for effective adaptation. Several traditional varieties and crops as well as cultivation practices are climate-smart. Practices like bunding, summer tillage, and mulching for water conservation, bushening in paddy, mixed cropping, etc., have been useful for risk adjustments. There is a need to first document as well as validate traditional technologies and grassroots innovations and then devise strategies for scaling up.

Strengthening Institutional Arrangements

Institutions play an important role in connecting people around the world to major technological changes to mitigate the effects of climate change. They are the main agents in transferring technologies and technical inputs to all the stakeholders, mainly the farmers. Many institutions are engaged in making technologies easily accessible, available, and affordable to farmers for better adaptation and sustainable livelihoods. However, there are important issues that need attention for improvement. Inadequacy of resources as well as competency among the change agency to cope with the challenges of handling the high-end technologies, mobilizing communities for conservation agriculture, and real-time advisory delivery is the limitation that plagues the technology transfer system severely. Though several initiatives have been taken but they are staggered and operate in silos and thus the momentum of change toward better adaptation is slow and weak. There are missing linkages and convergence of major institutions for the common goals. Failure of institutions to take cognizance of areas of high vulnerability and providing integrated solutions drawn from areas of better varieties, natural resource management, information technology, institutional innovations, etc., further retards the adaptation process. Convergence of institutions and integration of solutions could provide impetus to adaptation. A success story of climate change adaptation could be generated with water harvesting by constructing check bunds through the convergence of public and private institutions at Gaya in Bihar.

Institutional innovations like custom hiring centers have played a significant role in uptake of natural resource management (NRM) interventions, *viz.*, laser leveling, underground pipeline, irrigation with drip/sprinkler/rain gun irrigation systems (CESCRA, 2014) besides the adoption of conservation agriculture practices like zero-till wheat, direct seeded rice, and raised bed planting due to easy accessibility of costly and appropriate farm machineries.

Conclusion

There is a need to promote customized, need-based, farmers-centric, and sustainable mechanisms of adaptation in vulnerable communities. Emphasis has to be laid upon holistic development of people with proper knowledge and skill development, technological backstopping, and facilitation of institutional arrangements for village resource and custom hiring centers, ICT supported knowledge and advisory support system, as well as market linkage for effective adaptation and livelihood security among the households. Contingency plans with the integration of indigenous coping strategies and non-farm livelihood options should be made available. Climate change adaptation and natural disaster management have assumed great significance and to enhance awareness and preparedness it is essential to develop trained professionals. Integration of these subjects in formal as well as non-formal education systems will be immensely useful. Indian agriculture needs to be able to cope better with climate change and climate vulnerability if it has to increase its resilience to climate change. This requires states to adopt pioneering climate change adaptation projects, such as the National Innovations in Climate Resilient Agriculture (NICRA), which was launched in February 2011 by the Indian Council of Agricultural Research (ICAR).

References

Antwi, G. (2015). *The Influence of Agricultural Information Sources On The Practices And Livelihood Outcomes Of Cassava Farmers In Upper West Akim District.* Master of Philosophy Degree in Agricultural Extension, University of Ghana.

CESCRA. (2014). *NAIP Comp 3: Climate Change Adaptation (World Bank-GEF) (ICAR Code 303601). Strategies to Enhance Adaptive Capacity to Climate Change in Vulnerable Regions 2009–2014.* Center for Environment Science and Climate Resilient Agriculture (CESCRA), Indian Agricultural Research Institute. https://naip.icar.gov.in/downl oad/gef-iari.pdf.

Lal, S.P., Mahendra, A., and Singh, A. (2022). Dietary analysis of traditional food cultures in India: An overview of 2600 BCE to the 21st century. *Toros University Journal of Nutrition and Gastronomy-JFNG*, 1(1), 119–127.

Naresh, K.S., Bandyopadhyay, S.K., Padaria, R.N., Singh, A.K., Md. Rashid, Md. Wasim, Ranjeet Kaur, A., Swaroopa Rani, D.N, Panda, B.B., Garnayak, L.M., Suresh Prasad, M., Khanna, Sahoo, R.N., and Singh, V.V. (2014). *Climatic Risks and Strategizing Agricultural Adaptation in Climatically Challenged Regions.* IARI, New Delhi Publication.

Padaria, R.N., Rakshit, S., and Sadamate, V.V. (2017a). Aligning Agricultural Extension Strategies for Realizing the Targets of Sustainable Developmental Goals (SDGs). In *National Conference on Revisiting Agricultural Extension Strategies for Enhancing Food and Nutritional Security, Sustainable Livelihoods and Resilience to Climate Change- Towards Transforming Agriculture.* Organized by Sarvareddy Venkureddy Foundation for Development Participatory Rural Development Initiatives Society In collaboration with Professor Jayashankar Telangana State Agriculture University from 22–24, April 2017. http://www.prdis.org/pdfs/extncon_2017_ebook.pdf.

Padaria, R.N., Rakshit, S., Bandyopadhyay, S.K., and Pathak. H. (2017b). Climate resilient agriculture for food security: Farmers perspective and community initiatives. *Current Advances in Agricultural Sciences (An International Journal)*, 9(2), 123–131.

United Nations Publication. (2019). *Climate Change And Human Rights: Contributions by and for Latin America and the Caribbean (LC/TS. 2019/94).* Economic Commission for Latin America and the Caribbean/United Nations High Commissioner for Human Rights (ECLAC/OHCHR), Santiago.

Chapter 5

Integrated Pest Management: Need for Climate-Resilient Technologies

Md. Abbas Ahmad[*,‡], Deepak Kumar Mahanta[*],
Abdus Sattar[†], and Ratnesh Kumar Jha[†]

*Department of Entomology, P.G. College of Agriculture,
RPCAU, Pusa (Samastipur), Bihar, India*

†*Centre for Advanced Studies on Climate Change,
Dr. Rajendra Prasad Central Agricultural University,
Pusa (Samastipur), Bihar, India*
‡*abbas.ento@rpcau.ac.in*

Abstract

Climate change, one of today's most discussed concerns, poses serious challenges for farming on a global scale. Both agricultural output and insect pests are significantly impacted by shifts in precipitation patterns, rising CO_2 levels in the atmosphere, and rising temperatures. Pest insects may be impacted by climatic changes in a number of ways. They may lead to an increase in their geographic range, improved overwinter survival rates, more generations, altered plant-pest synchrony, altered interspecies interactions, higher risk of migratory pest invasion, more plant diseases transmitted by insects, and decreased efficiency of biological methods of pest control, particularly natural enemies. Consequently, there is great worry about the potential for agricultural economic losses and about the threat to people's access to food. Because climate change is a major factor in the growth and decline of pest populations, adaptive control strategies will be required. There are a variety of possible criteria that may be used to prioritize future research on the effects of climate change on agricultural insect pests. Climatic and pest population monitoring, Modified IPM strategies, and prediction tools based on modeling are a few examples.

Keywords: Global warming, Food security, Climate change, Pest management, Agriculture, Insect-pests.

Introduction

As the global human population has grown over time, so too have many aspects of people's day-to-day lives, technologies, cultures, economies, sciences, and food production. Agricultural revolutions, which have been impacted by the expansion of civilization, technology, and humankind in general, have brought about several significant changes in agricultural productivity. However, environmental changes and the remarkable population growth of the previous century have both contributed to concerns about the reliability of food supplies. By 2050, agricultural output around the world will likely need to double in order to meet the increased demand for food caused by a rising global population (Tilman *et al.*, 2011; Mahanta *et al.*, 2023). Increasing agricultural productivity, as opposed to clearing extra land area for crop production, has been demonstrated in several studies to be the most sustainable approach to guaranteeing food security (Godfray *et al.*, 2010). Rising global temperatures and atmospheric CO_2 levels, floods, droughts, severe storms, heat waves, and other extreme weather occurrences are among the topics of current scientific study in agronomy. A tendency toward reducing yield loss due to abiotic factors aforementioned has prompted increased focus on these issues in agricultural research. Where dry seasons currently act as a bottleneck to agricultural productivity, precipitation patterns that shift may have a greater impact on crop yield than temperature rises (Parry, 1990). Pests are a significant biotic element that are also influenced by climate change and changes in the weather. The dynamics of pest population, reproduction, survival, dissemination, and interactions with the environment and natural adversaries are all directly impacted by temperature change (Prakash *et al.*, 2014). As a result, it is crucial to keep an eye on pests' presence and abundance since the circumstances around their occurrence might rapidly alter. Insects, particularly invasive pest species, can pose a significant threat to agricultural production, and this chapter will focus on how climate change, specifically increasing temperatures and atmospheric CO_2 concentrations, along with variable precipitation patterns, affects insect biology and ecology. Possible solutions to the current issues in plant production are presented. These include updated integrated pest management (IPM) tactics, monitoring methods, modeling prediction tools, and the production of healthy food in an ecologically responsible way.

Constantly Evolving Climatic Situation

The climate has a significant role in determining the distribution and features of both managed and unmanaged systems, including those related to water resources, cryology, hydrology, freshwater and marine ecosystems, terrestrial ecosystems, agriculture, and forestry (Rosenzweig *et al.*, 2007). One definition of the phenomenon is the long-term change in weather conditions, such as humidity precipitation, and temperature. Global food production is in grave danger as a consequence of rising temperatures, climatic extremes, increasing CO_2 and other greenhouse gases (GHGs), as well as changed precipitation patterns (Fig. 1) (Shrestha, 2019). Today's globe is dealing

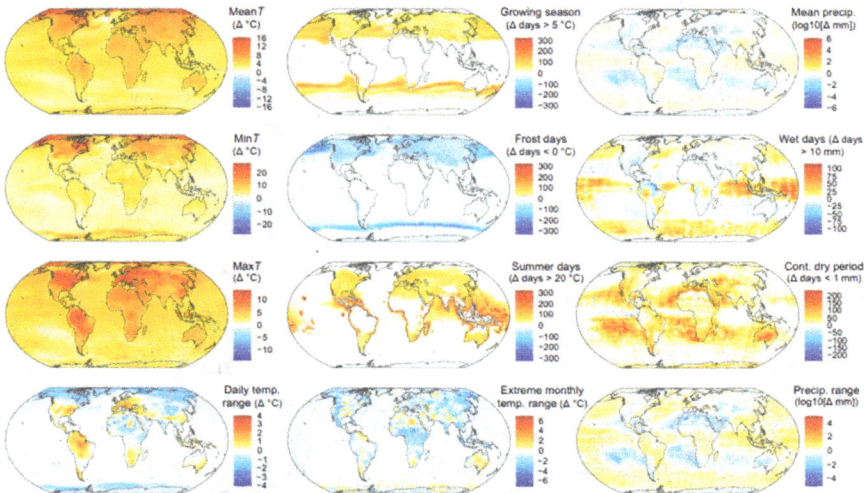

Fig. 1. Abiotic factors influencing the lives of insects. When comparing today's circumstances to those of the year 2100, each panel represents a different temperature. The Committee on the Environment's (COE) recommendations for climatic extremes (https://www.climdex.org) serve as the basis for the extreme indices used here.

with the significant issue of global warming. High rates of increase in both sea level and air temperature suggest that it has already reached unprecedented levels (Field *et al.*, 2014). According to the World Meteorological Organization (WMO), widespread industrialization has led to a one-degree global warming. According to the Intergovernmental Panel on Climate Change (IPCC), global temperatures have been rising steadily over the last three decades, with the 2000s being the warmest of all (Field *et al.*, 2014). The Earth may warm by 1.4–5.8°C over the next century, according to a variety of global climate models and development scenarios (Pachauari and Reisinger, 2007). Rising greenhouse gas concentrations in the atmosphere are the primary contributors to global warming. The most common human activities, such as burning fossil fuels and changing land use, are what produce the atmospheric gases methane (CH_4), carbon dioxide (CO_2), and nitrous oxide (N_2O) (Yoro and Daramola, 2020). When comparing the pre-industrial age to the previous two centuries of industrialization, the concentration of greenhouse gases has greatly grown (Rogelj *et al.*, 2018). The most significant and prevalent greenhouse gas is CO_2, which is also the most prevalent (Rosenzweig, 1989).

One of the most prevalent worldwide changes in the environment during the last 50 years has been the rise in atmospheric CO_2 (Prentice *et al.*, 2001). Compared to pre-industrial levels of 280 ppm, its atmospheric concentration has rapidly grown to 416 ppm, and it is estimated to become double in 2100 (http://www.esrl.noaa.gov/gmd/ccgg/trends/). Due to its significant absorption of some heat infrared radiation emitted from the Earth's surface, CO_2 is classified as a greenhouse gas. The fraction of thermal infrared radiation released down through the stratosphere and onto the ground increases with increasing atmospheric gas concentrations that absorb it (Mahlman, 1997). This means that there will be more energy available for latent and sensible heat flow and the long-wave balance at Earth's surface will be less negative. The amount of energy available for heat flux increases as a result, raising air temperature (Streck, 2005). Since the middle of the 20th century, changes in severe weather and climatic occurrences have been noted. The frequency of very heavy precipitation has risen in many locations, while the frequency of extremely cold temperature extremes has reduced. These are only a couple of changes that have been connected to human effects.

Excessive weather phenomena, such as heat waves and prolonged periods of extreme precipitation, are predicted to increase in intensity and frequency in certain regions (Field *et al.*, 2014). The likelihood of a fluctuating and uneven precipitation pattern is extremely high. The mean annual precipitation seems to be rising at the equatorial Pacific and the higher latitudes. While mean precipitation is predicted to rise in wet mid-latitude areas, it is likely to decrease in dry mid-latitude and subtropical regions. Extreme precipitation events are predicted to increase in frequency and intensity in the majority of tropical humid and mid-latitude regions (Field *et al.*, 2014). Numerous decisions have been taken by the IPCC and United Nations (UN) to cut GHG emissions, provide poor nations financial support, and boost the adaptive ability to address the difficulties presented by the adverse consequences of climate change.

Pest Insects and Climate Change

Global climate change has a major effect on agriculture and insect infestations. Alterations in weather patterns have direct and indirect effects on agricultural crops and the pests that attack them. Effects on pests' ability to develop, survive, reproduce, and spread may be seen immediately, whereas effects on natural enemies, adversaries, mutualists, and vectors can be seen indirectly via the networks formed among these insects and the pests they interact with (Prakash *et al.*, 2014). Insects are affected by their surroundings in terms of body temperature since they are poikilothermic. Since this is the case, it's safe to assume that temperature is the single most influential environmental factor for insect dispersion, behavior, reproduction, and development (Kocmánková *et al.*, 2010). Insect pest population dynamics, and hence the percentage of crop losses, are likely to be significantly influenced by the principal drivers of climate change (increasing atmospheric CO_2, increasing temperatures, and decreasing soil moisture) (Fig. 2) (Fand *et al.*, 2012). Because climate change is opening up previously uninhabited ecological niches, insect pests now have a greater possibility of flourishing, reproducing, and migrating from one region to another (FAO, 2020). The complex physiological effects of rising temperatures and CO_2 may have a significant effect on the interaction between insect pests and

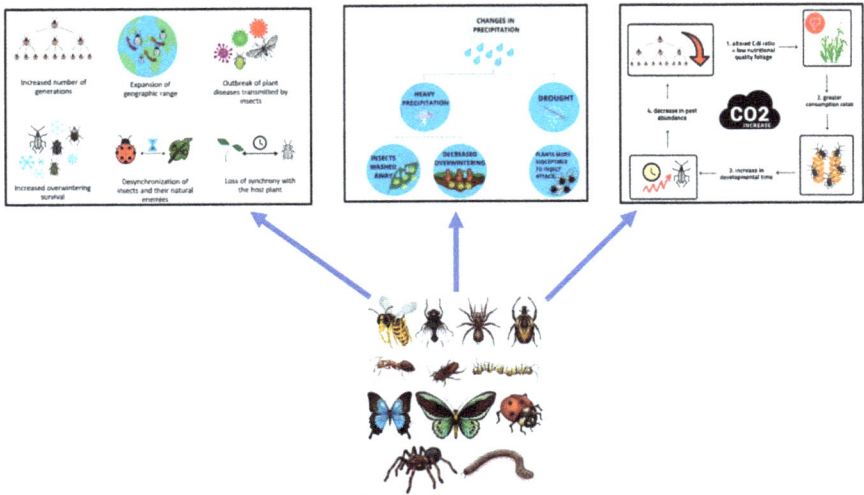

Fig. 2. Effect of (1) temperature, (2) precipitation, (3) and CO_2 on insects (Skendžič *et al.*, 2021).

agricultural crops (Roth and Lindroth, 1995). This means that in the next years, farmers will face new and severe pest concerns related to climate change. It is a global problem impacting all countries and regions because the spread of agricultural pests across borders poses a danger to food security (FAO, 2020).

Insect Pest Response to Increasing Temperature

Insects' physiology is highly temperature-dependent; an increase of 10°C often results in a doubling of their metabolic rate (Dukes *et al.*, 2009). Warmer temperatures here have been demonstrated in several studies to accelerate insect development, feeding, and migration. Population dynamics may be altered as a result of these factors, which may have an influence on reproductive success, longevity, population size, generation time, and dispersal (Bale *et al.*, 2002). The population of organisms that can't evolve to survive in warmer conditions dwindles while those that can do so thrive. A change in pest dynamics and population is possible only if the pests' metabolic rate, metamorphosis time, mobility, and availability of hosts all change in response to the temperature (Shrestha, 2019). Based on the current behavior and distribution of insects, it may be anticipated that rising temperatures and an increase in herbivory go hand in hand (DeLucia *et al.*, 2008). Warming temperatures may alter the behavior and location of insect pests, leading to altered population growth rates and perhaps increased herbivory (Deutsch *et al.*, 2018). Climate change is expected to slow the expansion of tropical insect populations since average global temperatures are already near to the optimum level for pest formation and growth while boosting the expansion of insect populations in temperate regions (Deutsch *et al.*, 2018). Scientists were able to lend credence to this theory by simulating the effects of several climate change scenarios on the production of wheat, rice, and maize — the three most important grain crops in the world. Wheat is normally grown in temperate climates, however, research suggests that the development of insect populations might be accelerated by rising temperatures. They foresee a slowing of the pace at which insect populations increase in tropical rice farms, but they predict contrasting responses to the same trend in temperate and tropical maize farms (Deutsch *et al.*, 2018).

For insects that spend most of their lives above ground, the effects of rising temperatures will be more severe than for those whose whole lives are spent below, due to the insulating properties of soil (Bale *et al.*, 2002). In warmer climates, for instance, aphids may be more vulnerable to predation because they are less likely to release the aphid alarm pheromone when threatened by parasitoids and insect predators (Awmack *et al.*, 1997). Environmental variables including precipitation, humidity, and temperature in general are the main regulators of whitefly populations. Whitefly population growth and high temperatures and humidity are positively correlated (Pathania *et al.*, 2020).

Future shifts in the dynamics of insect populations will depend on the pace at which global temperatures continue to climb over the next several years. Global temperatures are projected to increase by 1.8–4°C by the turn of the current century, according to climate models (Collins *et al.*, 2007). As the average temperature rises to a point where many insect pest species may thrive, the thermal constraints on their population dynamics are reduced, and the intensity of pest infestations is predicted to worsen under global warming scenarios (Deutsch *et al.*, 2008). Due to the restricted biological niche demands, physiological tolerances of insects, and varied effects of temperature on their phenology and life cycles, global warming may not necessarily lead to an increase in pest populations and, in turn, commercial crop losses (Lehmann *et al.*, 2020). Various analyses have shown that various insect pest species have varied responses to global warming (Lehmann *et al.*, 2020). Their research findings suggest that, in the majority of their insect case studies, rising temperatures enhance the severity of pest problems (Fig. 2). However, 59% of all the species examined had reactions that might lessen their negative effects, mostly via decreased physiological performance and range constriction. Another study of over 1,100 insect species found that 15–37% of those species would be extinct by 2050 due to climate change induced by global warming (Hance *et al.*, 2007).

Insect Pest Response to Increasing CO_2 Concentration

The range, population size, and overall output of plant-eating insects may change as a result of elevated CO_2 in the atmosphere. Fertility, consumption rates, growth, and population densities of insect pests may all be affected by these changes (Fuhrer, 2003). According to evidence now available, the impact of increased atmospheric CO_2 on herbivory is very unique to both specific insect species and insect pest-host plant systems (Fig. 2) (Coviella and Trumble, 1999). The impact of rising CO_2 levels on insect pests is strongly influenced by the plants that serve as their hosts. Wheat, rice, cotton, and other C_3 crops would be more impacted by rising CO_2 levels than C_4 crops (corn, sorghum, etc.). As a consequence, the distinct ways that high atmospheric CO_2 affects C3 and C4 plants may have asymmetric impacts on herbivory, and the way that insects that feed on C4 plants respond may be different from how they respond to C3 plants. While C4 plants are less sensitive to increased CO_2 and thus less likely to be impacted by changes in insect feeding behavior, C3 plants are more likely to be favorably influenced by elevated CO_2 and adversely affected by insect reaction (Lincoln *et al.*, 1984).

As was said before, higher CO_2 concentrations would likely affect plant physiology by increasing photosynthetic activity, leading to improved growth and increased plant production. Since this would affect the quantity and quality of leaves and plants, it would have a knock-on effect on insects. Chemical changes in leaves are a common

characteristic of plants in high CO_2 conditions, and these changes
may have an effect on the plant's nutritive value and its attractiveness to insect herbivores (Lincoln, 1993). Plants like this will
alter the carbon to nitrogen ratio and make themselves less palat-
able by storing sugars and starches in their leaves (Cotrufo *et al.*,
1998). Some insect pests, for whom nitrogen is a vital nutrient,
increase their plant-eating activity in response to elevated CO_2 levels (Bezemer *et al.*, 1998). Due to the increased plant tissue that
pests may consume in order to get the same caloric intake, plant
damage may increase. Insects that feed on leaves, such as caterpillars, miners, and chewers, often increase their consumption rates
as a kind of compensatory feeding when nitrogen levels decline as
predicted by CO_2 fertilizer (Hamilton *et al.*, 2005). Soybeans were
cultivated in an experiment by Hamilton *et al.* (2005) with high
atmospheric CO_2 levels. Early in the growing season, insects like the
Potato leafhopper (*Empoasca fabae* Harris), Western corn rootworm
(*Diabrotica virgifera* Le Conte), Japanese beetle (*Popilia japonica*
Newman), Mexican bean beetle (*Epilachna varivestis* Mulsant) and
caused 57% more damage to soybeans than they did to soybeans
grown in ambient atmospheric conditions. According to this study's
findings, the detected rise in the simple sugar content of soybean
leaves may have prompted compensatory insect feeding (Hamilton
et al., 2005). In these circumstances, insect herbivores often consume
more plant material, leading to increased plant damage (Lindroth
et al., 1993; Thomson *et al.*, 2010). Increased feeding rates may not
always make up for poor food quality, and eating plants grown in
high CO_2 environments may decrease the effectiveness of the arthropods that consume them (Zvereva and Kozlov, 2005). Pest feeding
habits may have a bearing on how plants respond to CO_2 fertilization.
A surge in the population may be seen in thrips and other whole-cell
feeders (Bezemer *et al.*, 1998). Insect pests that feed on phloem, such
as whiteflies and aphids, exhibit a dual reaction of faster population
growth rates and lower population densities (Sutherst *et al.*, 2011).
Reports on the impact of increasing CO_2 on sucking insects are con-
flicting, yet under certain circumstances, abundance and fertility may
rise (Hance *et al.*, 2007). In 2007, Stiling and Cornelissen carried out
a meta-analysis and assessed the research literature on the indirect
impacts of a CO_2 rise on herbivore life cycle characteristics. The findings of their study revealed that insect pests responded strongly to

elevated CO_2 levels compared to ambient CO_2; (I) consumption rates increased by about 17%; (II) pest abundance decreased by about 22%; (III) development time increased by about 4%; and (IV) relative growth rate decreased by about 9%. Additionally, chewers saw stronger impacts from the rise in atmospheric CO_2 compared to other feeding guilds, such as sap-sucking herbivores (e.g., aphids, leafhoppers, scale insects). As a result, it has been shown that despite the extensive research done to evaluate aphid reactions to rising atmospheric CO_2 levels, it is still impossible to anticipate future responses generally or to provide general guidelines for various aphid populations to climate change (Hullé *et al.*, 2010; Sun and Ge, 2011).

Insect Pest Response to Variable Precipitation Patterns

Modifications in precipitation's frequency, intensity, and quantity are all important barometers of climate change. For the most part, precipitation has become less frequent but more intense. Extreme weather events like floods and droughts are more likely to happen when this form of precipitation becomes the trend. Hibernating insect species are directly impacted by overlapping rainfall patterns (Fig. 2). In other words, heavy precipitation might lead to flooding and water stagnation over an extended period of time. This occurrence threatens the survival of insects and disrupts their seasonal hibernation. In addition, floods and severe rains may wash away bug eggs and larvae (Shrestha, 2019). Heavy rains may wash away small-bodied pests like jassids, aphids, whiteflies, mites, and others (Pathak *et al.*, 2012). On insect populations, variable rainfall may have a significant effect. For instance, Staley *et al.* (2007) investigated how soil-dwelling wireworms (*Agriotes lineatus* L.) responded to drought and increased summer rainfall in grassland plots. Because they cause significant damage to crops like potatoes, maize, sugar beet, and others planted on grassland plots, wireworms are predicted to become an even bigger issue as a result of climate change (Johnson *et al.*, 2008). According to Staley *et al.* (2007), higher summer rainfall episodes, as compared to ambient and drought circumstances, resulted in the fast proliferation of wireworm populations in the top section of the soil (Gregory *et al.*, 2009). Drought impacts herbivorous insects through a variety of processes; (I) Dry environments may provide favorable circumstances for the establishment and development of herbivorous insects; (II) Insects of some species are attracted to plants that are experiencing drought. By way of illustration, the acoustic ultrasonic emission created when water columns in the vascular tissue split apart or cavitate due to transpiration may be detected by destructive bark beetles (Scolytidae); (III) Due to decreased synthesis of secondary metabolites with a defensive role, plants under drought stress are more vulnerable to insect assault (Yihdego *et al.*, 2019).

Insect Pest Response to Extreme Drought Condition

Another climatic extreme that poses harm to insects is drought. The frequency and severity of extended (acute) droughts are growing in various locations, and they are accompanied by above-average temperatures, heat waves, and often fires (Dai, 2011; Williams *et al.*, 2022). Pulsed droughts, however, may last for a while before being momentarily ended by significant rainfall events (Harris *et al.*, 2018). Both forms of drought may directly affect an insect's physiological health or have an indirect impact on plant communities, which in turn affects insects that rely on them for food and shelter all the way to the top of the food chain (Gutbrodt *et al.*, 2011). The impact of drought stress on insects is complicated and influenced by a number of variables. For instance, insects that feed on trees may react to drought quite differently from insects that feed on forbs, sedges, and grasses, which are smaller plants (Gely *et al.*, 2021). Because tiny plants are more susceptible to water stress throughout the summer, drought spells may reduce the number of herbivorous insects on them. This will result in a scarcity of food supplies, which will have significant ramifications for population dynamics and interspecific interactions. Examples include a rise in competition among higher trophic levels for hosts or prey as a result of desiccation and subsequently loss of plant tissues. In contrast, insects that feed on trees are often "buffered" against drought because trees have far more biomass in their roots and shoots and can withstand more acute droughts than smaller plants.

Nevertheless, plants of any bulk may experience the effects of drought stress, which include chemical, physiological, and biological alterations (Anderegg *et al.*, 2015). Insect herbivores' growth and development may be impacted by changes in foliar and root concentrations of primary and secondary metabolites, such as defense allelochemicals and nutrients like amino acids and carbohydrates, during conditions of drought stress (Sconiers and Eubanks, 2017). Insect performance was connected to water stress from rising drought intensity in a recent study (Gely *et al.*, 2020). According to their predictions, various herbivore guilds would respond to drought stress in diverse but predictable ways, with the majority of guilds suffering unfavorable effects and several wood borers making a positive exception, at least in the near term. A few whole-forest drought modification studies have been conducted. Different feeding guilds of insects in a tropical rainforest in North Queensland, Australia, responded differently to an experimental drought (Gely, 2021). Significantly greater tree damage from wood borer occurred in the experimentally dry region than in the control area (Gely *et al.*, 2021). While certain ant species also depend on aphid honeydew, extrafloral nectaries are the source of nectar for many ant species in Australian rainforests. Drought-affected regions have few food sources, and stable isotope research suggests that many ant species are becoming more predatory (Gely, 2021), which will have an effect on the food webs in these woods. Due to certain insect eggs' water requirements for development, droughts might interfere with reproduction (Rohde *et al.*, 2017). The quality of floral rewards for pollinators may also alter due to dryness, which can reduce pollinator attractiveness and plant reproduction (Rering *et al.*, 2020).

Plant-insect communities may change even after a single severe drought. In 1995, the United Kingdom had a severe drought. As a result, the overall population of butterflies rose, but this was accompanied by significant changes in community composition, especially in more northerly, wetter places. Due to better options for recruitment from the bigger regional populations, specialist, vulnerable species have declined while generalist, widespread species have expanded. Communities had not reached equilibrium a year later (De Palma *et al.*, 2017), indicating that periodic droughts might increase the risk of extinction for both species and genetic diversity. In Arizona, butterflies were found to have a similar finding (Wagner and Balowitz, 2021). Single, severe droughts may lead to the genetic diversity being lost along with the final surviving ephemeral populations being extinct. Although the physiological and ecological processes behind reactions to severe drought are complex and poorly understood, the effects are becoming more obvious. The detrimental and protracted consequences of a recent megadrought in western North America on montane butterfly groups were comparable in scope to the cumulative effects of decades of habitat loss and degradation at lower altitudes (Halsch *et al.*, 2021). Even isolated indigenous species of dragonflies in the Cape Floristic Region, which experiences recurrent droughts, temporarily utilize manmade ponds to survive times of intense drought (Deacon *et al.*, 2019). Water beetles quickly left ponds when large droughts persisted, whereas dragonfly adults stayed devoted to the pond borders and foraged there until rains returned in the same area (Jooste *et al.*, 2020). These reactions show that freshwater insects might respond differently behaviorally to recurring droughts. Contrarily, it would be predicted that droughts would pose particular difficulties for less flier species and insects that have historically developed in perpetually damp to wet ecosystems, such as the fauna of cloud and rain forests (Janzen and Hallwachs, 2021; Wagner, 2020). Due to variations in plant quality and non-linear impacts up the food chain, climatic extremes like drought produce "winners" and "losers" among insects. However, when considered in the context of other anthropogenic pressures, the longer-term outlook for insects is unfavorable (Harvey *et al.*, 2020).

Insect Pest Response to Fire

The area, length, seasonality, and intensity of worldwide fire regimes have changed as a result of droughts and altered precipitation patterns (Jain *et al.*, 2021). Even while fire poses a deadly danger to many creatures, scientists are just now starting to see it as a crucial factor in climate change and a dynamic force that affects how species react to it (Nimmo *et al.*, 2021). Research findings on the effects of fire on insects vary depending on the weather, burn intensity, target taxa researched, and burn season (Banza *et al.*, 2021). The amount of research on invertebrate reactions to fire is constrained by arthropods' complicated life histories and responses that are often taxon-specific (Joern and Laws, 2013), which makes it difficult to propose effective conservation measures in response to significant fire outbreaks (Saunders *et al.*, 2021). We need more mechanistic research to improve our capacity to foresee the effects of shifting fire regimes. Periodic fires are necessary for the preservation of many insect species linked to early successional sequences and fire-adapted habitats. In fact, fires may attract a lot of wood-boring beetles as well as their predators. Some species, like the Araneae, are very sensitive to the effects of fire, while others, like the Coleoptera, are not. The effects of fire on arthropods range from negative to neutral to favorable (Kral *et al.*, 2017). The overall arthropod abundance may be strengthened by a robust regeneration of the herbaceous understory (Campbell *et al.*, 2007). The positive impacts of fires may be undone even for species that rely on them if fire regimes are drastically changed. For instance, localized reductions in species richness and/or abundance following fires have been noted in Australia (Andersen and Müller, 2000) and South Africa (Pryke and Samways, 2012a), though in South Africa at least, there can be rapid recovery as, for example, pollinators expand outward from fire refugia (Adedoja *et al.*, 2019). In any investigation, it's crucial to take into account not only that burning has different ecological effects depending on how sensitive or dependent an ecosystem is on fire, but also the different ways that fire is distributed spatially across these various landscapes in terms of size (10 vs. 10,000 ha^2), frequency, fuel loads within fire perimeters, and distance to refuges (Pryke and Samways, 2012a). For instance, it has been shown that certain plants with densely packed leaf bases may withstand strong flames while still providing shelter for insects and other arthropods (Brennan *et al.*, 2011). Further research is necessary to determine the significance of these refuges in the resilience of insect groups.

Global warming is anticipated to cause complicated changes in fire regimes, such as those brought on by phenological asynchronies in interactions between herbivores and enemies. The delayed post-fire recovery of parasitoids and the temporal variations of seasonal fires may have an impact on the accessibility of holometabolous hosts at certain life stages (Koltz *et al.*, 2018). Similarly, Dell *et al.* (2019) discovered that frequent fire caused a loss of specialized trophic interactions, which drove trophic webs toward generalization and increased the number of Orthoptera and Lepidoptera, which specialize in generalist feeding. Shorter burn cycles may thus lead to sporadic insect outbreaks. These (and other) instructions may have an impact on the structure and operation of communities if they are more effective at dispersing during severe wildfires and more quickly recolonize after the burn. For aquatic insects, particularly those that depend on terrestrial settings for a portion of their life cycle, fires may also have far-reaching effects. Some lentic taxa's eggs, for instance, are especially susceptible since they are dormant in the topsoil layers (Blanckenberg *et al.*, 2019). One effective way to monitor demographic changes is to make use of technology advancements in insect identification, such as eDNA metabarcoding, since there is limited information regarding the longer-term consequences of climatic extremes and associated events on insects (Jinbo *et al.*, 2011). This would be very helpful right away in the days, weeks, months, and years after a severe disaster, like a fire.

Expanding of the Distribution of Insects

In general, the following elements may impact where insect pests are found: crop distribution, Natural biogeography, agricultural methods (irrigation, monocultures, pesticides, fertilizers), commerce, cultural trends, and climate are only a few of the factors to consider (Ezcurra *et al.*, 1978). Low temperatures are often more important than high temperatures in defining an insect pest's geographic range, and climate change will have a substantial influence on this (Hill, 1987). Because of climate change and growing global commerce, which enables individuals to spread over the globe, many pest species are expanding their range. This sort of distribution change may have a significant impact on agricultural productivity in the case of agricultural insect pests (Meynard *et al.*, 2013). Species-specific climatic needs that are essential for their development, reproduction, survival, and growth highlight the geographic range and abundance of all creatures in nature. The survival, reproduction, and distribution of species in the future will be influenced by altered temperature and precipitation patterns as a result of the predicted changes in climate (Fand *et al.*, 2012). Due to insect pests' ability to adapt to new environments and the shift in where their host plants thrive, farmers will have to deal with a whole new set of challenging pest problems. In such situations, other parameters, such as soil characteristics and environmental structure, are crucial in addition to climate conditions appropriate for the specific crop (Laštuvka, 2010). As a result of global warming, a poleward shift in the distribution boundaries of pest species is anticipated (Bebber *et al.*, 2013). It is predicted that insect pest ranges would increase in altitude and more generations will occur in central Europe by the year 2055. As an example, the European corn borer has expanded its range in Europe by more than 1,000 km (Porter *et al.*, 1991). However, fewer generations were anticipated in southern Europe owing to global warming, which would have a detrimental impact on populations of this insect scourge. This suggests that distinct species are impacted by climate change in different ways (Shrestha, 2019). Lopez-Vaamonde *et al.* (2010) reported that 97 species of non-native Lepidoptera from 20 families have become established in Europe, whereas 88 species of European Lepidoptera from 25 families have expanded their distribution in the continent. Seventy-four percent of these have done so in the previous one hundred years. Parmesan *et al.* (1999) studied 35 non-migratory European butterfly species and found that over the 20th century, the geographic ranges of 63% of the species shifted between 35 and 240 km north, whereas just 3% shifted south. Warm air masses are expanding their range, allowing the Diamondback moth to establish itself on the Norwegian island of Svalbard in the Arctic Ocean, 800 km north of its traditional range limit in western Russia (Coulson *et al.*, 2002). An important cotton pest known as the pink bollworm is reportedly expanding its current range from the frost-free zones of California and southern Arizona into the cotton-growing areas of California Central (Gutierrez *et al.*, 2006). The olive fly distribution in North America and Europe will shift northward and retreat southward as a consequence of the effects of increased summer temperatures and milder winters on adult flies, as reported by Gutierrez *et al.* (2009). Because of the extreme cold in the far north, the olive fly habitat has been reduced to the deserts of central and southern California and Arizona. It is anticipated that as high summer temperatures grow more unfavorable, climate change may further restrict its occurrence in many Californian locales. However, the environment along the coast of California is thought to be more conducive to their growth. Northern Italy's olive and flies are unable to survive the cold winters because of the lack of a suitable climate, but this is expected to change as formerly unfavorable sites become favorable as a consequence of global warming (Gutierrez *et al.*, 2009). On the other hand, variations in the pattern of frost are a factor in the proliferation of pest insects that are sensitive to the cold (Fleming and Volney, 1995). Longer mild periods lengthen the duration and severity of insect epizootics because the frequency of spring frosts declines with rising temperature (Raza *et al.*, 2014). Although crop growers may theoretically benefit from earlier seeding, this practice also has drawbacks. Insect pests might begin feeding sooner, doing greater harm to the plants they infest. There may be more generations of insects throughout the regular growth season (Raza *et al.*, 2014). More insects will be able to survive the colder winters at higher elevations, which might lead to the spread of these insects' range (Patterson *et al.*, 1999; Pareek *et al.*, 2017).

Increased Number of Generations

The most significant environmental component for insects, primarily determining their phenology, is temperature, as was already noted. According to the ambient energy concept, growth and reproduction are boosted by hot weather. Consequently, rising temperatures or global warming cause larger populations, which may result in more species in dynamic equilibrium (Menéndez *et al.*, 2007; Menéndez, 2007). With this, many insect species may increase their reproductive rates within a limited range, leading to more generations and more agricultural harm in the event of global warming (Yamamura and Kiritani, 1998). Among the many species characteristics and environmental factors used to link climate change to phenological changes in growing degree days (GDD). Using the daily cumulative sum between a maximum and minimum temperature (D_{max} and D_{min}) threshold, the GDD is a metric of heat buildup that is determined yearly. In agriculture, GDD has been used to forecast insect and plant phenology (Cayton *et al.*, 2015). Univoltine and multivoltine temperate species will be impacted differently and to varying degrees by future temperature changes. Increased temperatures should allow insects like certain lepidopteran species and aphids (multivoltine) like the big cabbage white butterfly (*Pieris brassicae L.*) to develop at a faster rate, leading to numerous generations within a year (Bale *et al.*, 2002; Pollard and Yates, 1993). In general, species with yearly life cycles grow faster than those with longer life cycles (Bale *et al.*, 2002). It has been calculated that a 2°C rise in temperature might lead to one to five more life cycles every year using a variety of models (Yamamura and Kiritani, 1998). Aphids are the most notable instances in this respect, and because of their low developmental threshold and rapid generation rate, they may be anticipated to generate four to five more generations per year. Aphids may consequently be very adept at temperature change indicators (Menéndez, 2007). As a result of shorter nymphal and larval phases (during which they are severely endangered by predators) and earlier adulthood (as a result of higher temperatures throughout their development) (Bernays, 1997), some species are able to reproduce more successfully (Menéndez, 2007).

An earlier adult emergence and a lengthening of flight are two adaptations that insects should have in response to a rise in temperature (Menéndez, 2007). One possible reason for the voltinism variances is an earlier start to the flight phase, which would allow for the development of an additional generation (Altermatt, 2010). It's possible that the first generation of insects will start reproducing earlier since they can take to the air earlier in the growing season. Additionally, because of quicker larval development and growth brought on by warmer temperatures, more members of the next generation may develop while temperature and photoperiod circumstances are still favorable, enabling them to develop in the current season rather than halting to develop as larvae (Altermatt, 2010). Pheromone, suction, or light traps may be used to record the time of adult emergence. The timing of the appearance of insect pests alters as a result of climate change, according to long-term data analysis on insect phenology (Pathak et al., 2012). Analyzing data from suction traps, researchers found that for every 1°C increase in mean temperature in January and February, the spring flight of the Potato aphid began two weeks earlier (Shrestha, 2019). Population densities rebound from the winter in a wide range, from very low (very cold winter) to very high (very long) exposure (mild winter) (Harrington et al., 2007). Rothamsted Research, Harpenden, UK, conducted a 50-year study of the period of the first migratory individuals of aphid captured in a suction trap every year, and they found a substantial correlation with winter mean temperatures in January and February (Bale and Hayward, 2010). Another excellent example of phenological fluctuations is seen in the members of the order Lepidoptera. In the UK, where 35 species of butterflies have been seen, 26 of those species have undergone some kind of butterfly evolution that has resulted in an earlier adult appearance (Menéndez, 2007; Roy and Sparks, 2000). There has been a 1–7 week shift in the earliest arrival of 17 species in Spain during the last 15 years (Menéndez, 2007; Stefanescu *et al.*, 2003). In Spain, the early advent of the European grapevine moth led to an increase in voltinism. Although this pest is generally trivoltine in Mediterranean latitudes, it occasionally has a fourth additional flight due to its tendency to emerge early in the spring. This may be a consequence of climate change (Martín-Vertedor *et al.*, 2010). Some species of central European Lepidoptera that formerly had univoltine or bivoltine life cycles have recently shifted to bivoltine or multivoltine patterns (Altermatt, 2010). The abundance of second or later generations is anticipated to grow in partially bivoltine or multivoltine species (Altermatt, 2010; Roff, 1980).

Given the complexity of insect pests, it is difficult to pinpoint how climate change may affect specific species, climatic circumstances, and interconnected ecosystems (Bale *et al.*, 2002). However, a conceptual groundwork for how these specific changes may arise in other insect species could be laid by an exact assessment of the link between insect traits and climate change, such as phenology and voltinim shifts for a key insect nuisance species (Tobin *et al.*, 2008). Insects are among the creatures that react to global warming because their remarkable adaptation to environmental change is confirmed by the observed alterations in voltinism (Yamamura and Kiritani, 1998).

An Increase in the Frequency of Plant Diseases Spread by Insect Vectors

Insects are key vectors for numerous plant diseases such as viruses (Bhoi *et al.*, 2022), phytoplasmas, and bacteria (Boland *et al.*, 2004). To a large extent, viruses are responsible for the spread of many diseases that affect plants used in the production of food for human consumption (Mahanta *et al.*, 2022). The annual economic cost of these diseases is projected to be more than $30 billion (Sastry and Zitter, 2014). Viruses are immobile outside of their vector or host insect and hence rely largely on them for transmission and dissemination.

While certain viruses and vectors are able to infect a wide variety of hosts, others are highly specialized in a certain mode of transmission. Vectors, their host plants, and the environmental circumstances in which they thrive all have a role in the persistence, spread, and prevalence of viruses (Hull, 2014; Trebicki, 2020). An important factor in the spread of plant viruses might be altered by climate change (Trebicki et al., 2016). Viruses that replicate only on one strand of DNA or on a messenger RNA are the most common kind of viruses that infect crops grown for human use. Insect vectors with piercing and sucking mouthparts are most often used for transfer from one host to another (Canto et al., 2009). Earlier, we covered how climate change is already having an impact on a wide range of insect pests, some of which also act as virus vectors. Because climate has a direct impact on insect physiology, phenology, and so on, it may have an indirect impact on the viruses they spread. This effect might have a favorable, negative, or neutral impact on the introduction and progression of viral infections in agricultural production (Trebicki, 2020).

The extension of their geographic range and the growth of their insect vector populations are two ways in which climate change may facilitate the establishment of previously unknown insect-transmitted plant diseases (Sharma et al., 2005; Sharma, 2014). The aphid, leafhopper, and whitefly families are the most important vectors of viral infections within this order (Nault, 1997). Aphids constitute the biggest group of these vectors, spreading more than 275 different types of viruses, and aphid species are able to spread certain plant viruses (Bhoi et al., 2022). While whiteflies are confined to warmer climates and do not flourish in temperate locations, aphids are important viral vectors in temperate sections of the globe and are found in crops cultivated in greenhouses (Canto et al., 2009). Aphids and whiteflies are especially vulnerable to the effects of climate change due to their rapid growth and strong reproductive potential (Hance et al., 2007). Climate change may also have an impact on the ability of viral vectors to migrate and their ability to disperse across large distances. Aphids are capable of traveling great distances when they are launched higher by favorable temperature conditions, where they are then exposed to horizontal translocation caused by atmospheric air movements (Fereres et al., 2017). This long-distance transmission has been associated with severe virus outbreaks spread by aphids from the Great Plains of North America in the south to Minnesota corn-growing regions by highly persistent low-pressure winds (Zeyen *et al.*, 1987).

Warmer temperatures in Northern Europe at the beginning of the growing season have also been linked to a higher incidence of viral infections in potatoes since aphids-the principal vectors of these pathogens-had already established an early foothold there (Fand and Kamble, 2012; Robert *et al.*, 2000). The time and quantity of the inoculum have a significant impact on the severity of viral infections. Influenced by the insect vectors' and their (alternative) host plants' overwintering, the quantity of viral inoculum (Irwin *et al.*, 2000). Milder winters are anticipated to boost aphid survival rates, and warmer spring and summer temperatures will hasten aphid growth and reproduction. The net result is an increase in the frequency of viral illness transmission and dissemination (Alonso-Prados *et al.*, 2003).

The Poaceae family of plants suffers from a disease called barley yellow dwarf virus (BYDV), which is spread by a number of aphid vectors. Based on extensive observation, the minimum temperature required for the migration of the principal BYDV vector, the Bird cherry oat aphid (*Rhopalosiphum padi* L.), is 8°C in Central Europe. Additionally, population growth in autumn depends on precipitation patterns and particularly cold winter temperatures, whereas population growth in fall depends on temperatures in autumn (Jarošová *et al.*, 2019).

The danger of viral transmission in winter crops like winter wheat and winter barley increases due to warmer weather in central and northern Europe throughout the fall and winter (Roos *et al.*, 2011). When it's hot outside and there isn't much rain, viruses and the insects that carry them face a variety of difficulties. Aphid survival declines above 36°C during the hottest summer months, limiting the BYDV spread (Parry *et al.*, 2012).

The two most significant viral vectors among whiteflies are the greenhouse whitefly and the silver leaf whitefly. Whitefly thrives best in conditions with moderate precipitation and high temperatures, which results in population growth (Seruwagi *et al.*, 2004). Environments with established irrigation systems and hot, dry weather are favorable for *B. tabaci*. Given their rapid rate of reproduction, summer is the best period for big populations to form. The same circumstances might accelerate the virus's development, giving rise to variants that are more effective and have a wider host range, higher transmission efficiency, and bigger viral reservoirs in crops. Climate change projections suggest that extreme winds and greater cyclonic activity would occur in the tropics, which might aid in the spread of *B. tabaci*. Drought may reduce its chance of survival, obstruct its growth, limit population size, and prevent population spread (Sutherst *et al.*, 2011). According to climate simulations, many more places throughout the globe will be ideal for growing outdoor tomatoes under four distinct climatic scenarios that take temperature, humidity, and atmospheric CO_2 levels into account. These areas could potentially develop into favorable environments for the growth of *B. tabaci* populations, leading to a rise in the incidence of tomato yellow leaf curl virus, a very destructive disease (TYLCV) (Ramos *et al.*, 2019).

Grapevine yellows are a disease of grapevines connected to phytoplasmas. Because their associated insect vectors have varied life cycles, they exhibit significant epidemiological disparities (Boudon-Padieu, 2005). One of the most important grapevine diseases in Europe is Flavescence dorée (Boudon-Padieu, 2003), and its vector is the American grapevine leafhopper (Falzoi *et al.*, 2014). *S. titanus* is extending its range northward as growing season average temperatures rise (Boudon-Padieu and Maixner, 2007). Short summers are thought to prevent the spread of *S. titanus* in the north because the insect cannot complete its life cycle (Boudon-Padieu, 1932; Mirutenko *et al.*, 2018). Warmer and longer summers as a consequence of climate change, however, could aid the growth of *S. titanus* in northern vineyards like those in Germany. *S. titanus* is extensively distributed across several European regions that cultivate vines. *S. titanus* was discovered by scientists in Ukraine, which is now the northernmost point of its range in Europe. However, climate change toward the southern limit of its existing distribution may cause insect populations in places like southern Italy to dwindle or become extinct (Boudon-Padieu, 1932). It is anticipated that newly established insect-transmitted plant diseases may proliferate due to climate change. Therefore, having diagnostic equipment and qualified workers is crucial for finding novel diseases.

Increased Threat from Alien Insect Species

The term "invasive alien species" (IAS) refers to taxa that are introduced either knowingly (e.g., as food, crops, ornaments, pets, or livestock) or unknowingly (e.g., as a result of human activity beyond their native environment) (Shine *et al.*, 2000). Typically, pests of agriculture, forestry, stored goods, homes, structures, and invasive insects may also serve as carriers of a number of illnesses and parasites (Ward and Masters, 2007). International travel, the global commerce system, and agriculture have all contributed to an exponential increase in the spread of species during the last millennium to areas outside of their native range (Ricciardi, 2013). Invasive alien species are considered the biggest danger to biodiversity (CBD, 2018) worldwide, posing substantial costs to agriculture, forestry, and aquatic ecosystems, according to the Convention on Biological Diversity (Shrestha, 2019). It is a widely held belief that only a tiny percentage of newly imported IAS establish themselves and that only a small percentage of these species proliferate and become commercial pests. According to the "rule of 10," which is often used, around 1 in 10 invasive species escape into the ecosystem, 1 in 10 of these introduced species settle in the environment, and 1 in 10 of these established species turn into economic pests (Vander Zanden, 2005).

Multiple authors of recent research have concluded that invasive insect pest species would likely increase their range, population density, and voltinism in response to future climate change (Walther *et al.*, 2009; Hill *et al.*, 2016), which could have immediate and potentially negative effects on the sustainability of agricultural production (Ziska *et al.*, 2011). It is crucial to note that biological invasion is not primarily caused by climatic change. Aliens insects must successfully colonize a new area, endure the environmental challenges, and flourish there before they may become invasive. Climate change may or may not have an impact on the elements of this invasive route. The seasonal circumstances for these species' development, growth, and survival in a new environment are determined by climate, which also influences landscape characteristics (Masters and Norgrove, 2010). These habitats may have been unsuitable in the past, and a physical barrier, such as sea or mountain ranges, may have prevented dispersion to appropriate, far-off habitats (Vanhanen, 2008). Temperature increases will have a significant influence on ecosystems and the animals that inhabit them since all biological systems have thermal limitations.

However, it is already evident that not all insect species, both native and invasive, will thrive from the new, higher temperatures as a result of global warming (Masters and Norgrove, 2010). A series of activities, including the transportation, introduction, establishment, and spread of invasive alien insects, make up the process of insect invasion (Ricciardi, 2013). The remaining phases of the invasion process may be favorably or adversely impacted by the current temperature and climatic change after a new species has arrived in a new habitat (Tobin *et al.*, 2014). Invasive insect invasion and transportation may be significantly impacted by climate change. Extreme weather conditions, such as storms, hurricanes, strong winds, currents, and waves, might move pests to new regions where they can find environmental conditions that are conducive to their development (FAO, 2020). For instance, during the 2005 hurricane season, the Cactus Moth (*Cactoblastis cactorum* Berg) was transported from the Caribbean islands to Mexico, where it caused a serious ecological and financial danger to more than 104 prickly pear species, 38 of which are indigenous (Hill *et al.*, 2007; Burgiel and Muir, 2010). Certain paths favor the introduction of particular alien insect species, and some insect species are more likely than others to be introduced and dispersed to new geographic areas (Kiritani and Yamamura, 2003). Alternatively called as an "introduction effort," propagule pressure (Vanhanen, 2008) is the quantity of insect individuals that are arriving (Lockwood *et al.*, 2005).

The frequency and quantity of individuals entering a new environment determine the propagule pressure (Ward and Masters, 2007). In general, the more people who are brought into a region, the higher the likelihood that they will establish (Lockwood *et al.*, 2005). A species' propagule or propagules must first enter a route for transportation, endure the journey, leave the pathway successfully, and then create an initial population that may or may not expand and become invasive (Simberloff, 2009). Propagule pressure is correlated with the volume of plant commerce, the risk that these plants will be used to transport foreign insects, and the likelihood that these insects will sneak by border inspections in plant commodities (Bacon *et al.*, 2014). One of the most recent instances of such an introduction route is the invasion of North and South America as well as Europe by the extremely polyphagous and invasive spotted wing drosophila (*Drosophila suzukii* Matsamura). The trade of fresh fruit is assumed to be the source of the introduction, with the first propagules developing undetected in the egg or larval stage in significant amounts of fresh fruit trafficked across Southeast Asia (Rota-Stabelli *et al.*, 2013; Cini *et al.*, 2014).

Insects from other countries may be able to find more places that suit them than local ones can, since invading species tend to have a larger bioclimatic range (Walther *et al.*, 2009). It is well known that many insect species are quite susceptible to climate change. Due to the fact that the majority of their physiological activities rely on temperature, they are sensitive (Vermeij, 1996). A major factor in the propagation of many invasive species is plasticity. Because it is a personal attribute, plasticity is often hailed as a responsive mechanism that enables creatures to adapt to new environmental situations in a world that is changing quickly (also known as "plastic rescue") (Snell-Rood *et al.*, 2018; Fox *et al.*, 2019).

Adaptations may manifest as phenotypic, behavioral, developmental, or physiological features. Different environmental factors (such as temperature, humidity, and photoperiod), the availability of a particular food, or pressure from predators or rivals may cause physiological or behavioral plasticity (Abram *et al.*, 2017; Sgro *et al.*, 2016). Finding host plant species when invading new ecosystems is one example of how behavioral responses may be adaptive and increase fitness. Foraging insects have flexible reactions to changing settings, one of which is to alter or broaden their food choices. The food breath of certain species, like *D. suzukii*, which exhibits extraordinary versatility in its diet choice with more than 30 plant species, is likely the most crucial characteristic influencing the success of its invasion (Poyet *et al.*, 2015). Components of many systems, such as plastic responses to photoperiod in connection to climatic change, are involved in the evolution of numerous features (Snell-Rood *et al.*, 2018). According to Snell-Rood *et al.* (2018), general mechanisms that develop as a result of selective processes within an individual are very likely to result in survival in unfamiliar environments, particularly when conditions differ markedly from those in the native environment, such as significant temperature changes.

Thermal adaptation in ectotherms, such as insects, may happen, for instance, via behavioral characteristics that regulate energy consumption (Chevin *et al.*, 2010).

Strategies for Pest Management in a Changing Climate conditions: Adaptation and Mitigation

Implementing current risk management techniques and lowering the potential risk from climate change effects may be seen as a continuous process of climate change adaptation (Howden *et al.*, 2007). Pest infestations are anticipated to become more unpredictable and spread over more areas due to climate change. The connections between insects and plants in ecosystems are yet unknown, which adds to the ambiguity around how climate change may directly affect agricultural harvests (Gregory *et al.*, 2009). Numerous biological, economic, and social elements will affect how adaptable agricultural production systems are. The physical, social, and economic resources of local communities will determine their capacity to modify their pest control methods (Sutherst *et al.*, 2011). Uncertainties and the frequency of both new and old pests' occurrences will rise as a result of climate change and the acceleration of global commerce. Therefore, improving quick adaptation to shocks and climate changes will be even more crucial (Barzman *et al.*, 2015). To lessen the chances of new illnesses and pests spreading and to lessen the effects of current pests, potential adaptation measures have been developed. Modified IPM techniques, climate monitoring, insect pest population monitoring, and the use of modeling prediction tools are the most often stated options (Fig. 3) (Raza *et al.*, 2014).

Modified IPM Practices

IPM refers to the prevention and management of plant diseases, weeds, and harmful species of phytophagous organisms (mostly insects and mites). Sustainable agriculture places a premium on using preventive or indirect means of plant protection before resorting to control or direct measures. When determining whether or not to implement controls, it is essential to apply cutting-edge methods,

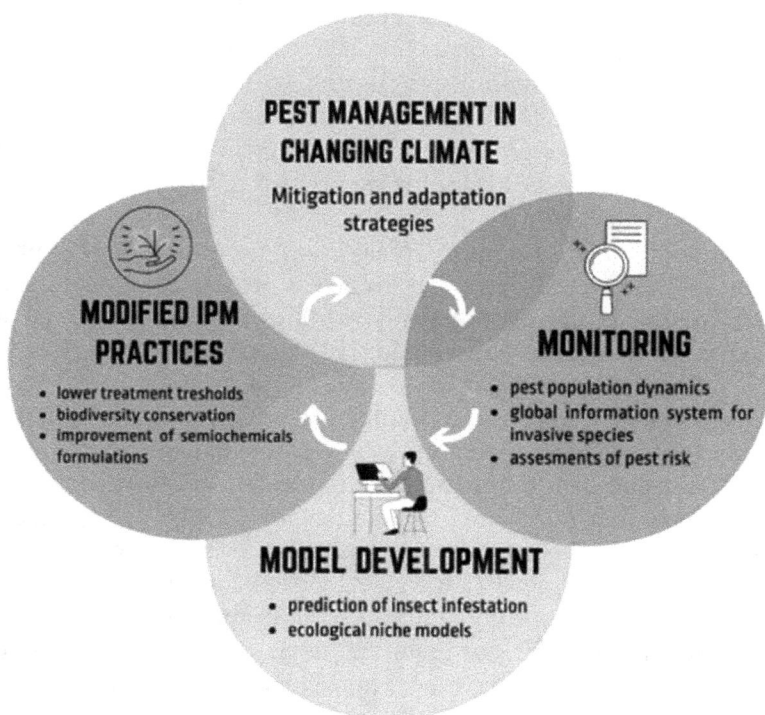

Fig. 3. Methods for controlling pests that may help us adjust to our changing climate (Skendžič *et al.*, 2021).

such as scientifically validated forecasting strategies and thresholds (Fig. 3). Direct pest control tools are the last resort when indirect methods fail to prevent economically intolerable losses (Boller *et al.*, 2004). The FAO suggests a two-fold approach that involves both global and regional action and, most importantly, a sizable investment in enhancing current early detection and control mechanisms. To stop the spread, new agricultural techniques must be developed, new crop species must be introduced, and integrated pest control methods must be used (Gomez-Zavaglia, 2020).

IPM techniques are primarily created by producers and academics to reduce adverse environmental effects while optimizing agricultural yields and financial returns (FAO, 2021). Numerous writers have emphasized the challenge of pest control in a new setting with a changing climate and the necessity to reevaluate current preventative agriculture practices and IPM tactics to develop diverse agroecosystems that are robust enough to handle weather unpredictability (Barzman *et al.*, 2015). To address the significant effects of global warming, researchers and producers will need to alter many of these well-designed IPM strategies, according to recent predictions (Barzman *et al.*, 2015).

Multiple IPM efforts have focused on determining the precise number of insect pests that may coexist with a crop before production losses become economically significant (also known as "economic" or "intervention") threshold. IPM has traditionally developed in the realm of pest control, where the use of predetermined thresholds has produced positive outcomes. Even while intervention levels are crucial to IPM, they are not always applicable, adequate, or even feasible. Thresholds are not used when decision support systems are unavailable or not suitable (Barzman *et al.*, 2015). It is crucial to comprehend how the environment influences the growth of plants and pests since this knowledge enables agricultural advisers to adapt to climate change. Recommendations for crop protection are affected by environmental conditions including drought stress. The economic threshold may readily be lowered when a crop is under drought stress because it is less able to handle the extra stress brought on by herbivorous insects (Lamichhane *et al.*, 2015). Because insects grow more quickly at rising temperatures, populations grow more quickly and agricultural damage happens sooner than presently anticipated. To avoid unacceptably high yield losses, treatment limits depending on the number of insects per plant must be decreased (Trumble and Butler, 2009). To lessen the effects of agricultural pests on crops in a changing environment, modified farming methods and adaptive management measures are required. To reduce vulnerability to pest outbreaks, these may include (a) planting several crop kinds; (b) planting at various times of the year; and (c) enhancing biodiversity at field edges to enhance the number of natural enemies (Andrew and Hill, 2017).

Pheromones and allelochemicals are used by insects in a critically important process: perceiving their environment. Numerous IPM strategies rely on them, including biological control, mating disruption, push-pull methods, monitoring, and trapping (Heuskin *et al.*, 2011). It is expected that as the climate warms and microclimates become more variable, the current methods of using pheromones and allelochemicals will be less effective, and they may need a synergist or other adjuvant to minimize their volatility under high-temperature situations (Andrew and Hill, 2017). Additionally, a few biopesticides based on enthomopathogenic bacteria, nematodes, fungi, viruses, and bacterial strains are quite vulnerable to environmental changes. Some of these control strategies may become less effective when temperature and relative humidity rise, and synthetic pesticides are predicted to have a similar impact (Nihal, 2020). In this situation, the emphasis should be on the creation of fresh approaches to pest control as well as potential advancements in pesticide formulations, repellents, and attractants. For instance, Wenda-Piesik *et al.* (2015) looked at the behavior of the confused flour beetle (*Tribolium confusum* Du Val) in relation to various concentrations of ecologically favorable volatile organic compounds (VOC) and their attractive and repulsive qualities. They were able to establish that individuals of the indicated species were strongly repulsed by the greatest concentration of applied VOC. This study may act as a springboard for the creation of brand-new, ecologically responsible pest management solutions. Understanding how global warming affects the effectiveness of several synthetic insecticides, their longevity in nature, and the emergence of pesticide resistance in pest populations is urgently needed (Vadez *et al.*, 2012). It seems necessary to think about using efficient biological control agents or introducing new varieties of insect-pest-resistant crops that were generated via conventional genetic breeding or genetic engineering (Gomez-Zavaglia *et al.*, 2020).

Monitoring of Distribution and Abundance

To determine whether the population dynamics of insect pest species are being affected by climate change, long-term data are essential (Yamamura *et al.*, 2006). Without these essential baseline data, analyzing population shifts in pests in response to changing climate regimes and forecasting future population dynamics is very challenging (Fig. 3). Long-term monitoring of pest numbers and behavior, particularly in places vulnerable to climate change, may give some of the first clues of biological responses to climate change (Heeb *et al.*, 2019). Trends in the regional dispersal of vectors, diseases, and host populations should be monitored, just as closely as their local dynamics. New invasive species are being introduced into different parts of the world because of climate change. Invasive species must be effectively monitored and managed to avoid becoming a financial burden in new geographic areas (Hellmann *et al.*, 2008). Therefore, both pest control and biosecurity will need adaptive responses.

Because of the effects of global warming, it may be required to step up the use of various pest management strategies, including surveillance and forecasting, as well as chemical, physical, and biological methods (Heeb *et al.*, 2019). Due to the transboundary character of many insect pests, effective monitoring and risk assessment need a worldwide management strategy. Important data on insects, invasive alien species, diseases, and ecological conditions, including meteorological information, must be shared throughout regions, hence a global system for this purpose is required. Consequently, it is critical to strengthen international collaboration, including that of national, regional, and global organizations (Perrings *et al.*, 2010). Rapid eradication and entry point surveillance, as shown by the US Department of Agriculture (USDA). Both the European and Mediterranean Plant Protection Organization's (EPPO) Early Warning and Rapid Response Program and its Early Warning and Information System for IAS will continue to play an important role in the fight against invasive species (Joyce *et al.*, 2013). In addition, by keeping an eye on the weather and pests and armed with knowledge of climate and pest risk prediction, farmers may take preventative measures to decrease the growth and spread of predicted pest difficulties (Heeb *et al.*, 2019).

Conclusions

Although there are still a lot of unanswered questions about climate change, it is commonly acknowledged that it has a significant impact on both the development of agricultural plants and the insect pests connected to them. Small-scale climatic variability, such as temperature rise, increase in atmospheric CO_2, shifting precipitation patterns, relative humidity, and other variables are some of the uncertainties surrounding various elements of climate change that are pertinent to insect pests. It is projected that various insect species would react differently to global warming in different regions of the globe due to the huge variety of insect species, their host plants, and global climatic variability. Climate change has complicated consequences on insects since it affects their range, variety, abundance, development, growth, and phenology while favoring some and hindering others. Even more concerning is the widespread belief that new and different kinds of insect pests will become a problem as a result of this. Insects would undoubtedly start showing up in new places (mostly northward). Some pests will become more numerous as a result of higher overwintering survival rates and the capacity to produce additional generations. There would most certainly be an increase in plant illnesses spread by insects as well as invasive pest species that may colonize new places more easily. The diminished efficacy of biological control agents, or natural enemies, is another unfavorable impact that climate change could have, and this might provide a significant challenge for future pest management operations. We run a high danger of suffering substantial financial losses and a threat to human food security if climate change variables create favorable circumstances for insect infestation and crop destruction. To solve this issue, a proactive and scientific strategy will be needed. Planning and developing adaptation and mitigation methods, such as improved IPM techniques, monitoring of pests and the climate, and the use of modeling tools, are thus very important.

References

Abram, P.K., Boivin, G., Moiroux, J., and Brodeur, J. (2017). Behavioural effects of temperature on ectothermic animals: Unifying thermal physiology and behavioural plasticity. *Biological Reviews of the Cambridge Philosophical Society*, 92(4), 1859–1876. https://doi.org/10.1111/brv.12312.

Adedoja, O., Dormann, C.F., Kehinde, T., and Samways, M.J. (2019). Refuges from fire maintain pollinator–plant interaction networks. *Ecology and Evolution*, 9(10), 5777–5786. https://doi.org/10.1002/ece3.5161.

Alonso-Prados, J.L., Luis-Arteaga, M., Alvarez, J.M., Moriones, E., Batlle, A., Laviña, A., García-Arenal, F., and Fraile, A. (2003). Epidemics of aphid-transmitted viruses in melon crops in Spain. *European Journal* of Plant Pathology, 109(2), 129–138. https://doi.org/10.1023/A:10225 98417979.

Altermatt, F. (2010). Climatic warming increases voltinism in European butterflies and moths. Proceedings Biological Sciences, 277(1685), 1281–1287. https://doi.org/10.1098/rspb.2009.1910.

Anderegg, W.R., Hicke, J.A., Fisher, R.A., Allen, C.D., Aukema, J., Bentz, B., Hood, S., Lichstein, J.W., Macalady, A.K., McDowell, N., Pan, Y., Raffa, K., Sala, A., Shaw, J.D., Stephenson, N.L., Tague, C., and Zeppel, M. (2015). Tree mortality from drought, insects, and their interactions in a changing climate. New Phytologist, 208(3), 674–683. https://doi.org/10.1111/nph.13477.

Andersen, A.N. and Müller, W.J. (2000). Arthropod responses to experimental fire regimes in an Australian tropical savannah: Ordinal-level analysis. Austral Ecology, 25(2), 199–209. https://doi.org/10.1046/j.14 42-9993.2000.01038.x.

Andrew, N.R. and Hill, S.J. (2017). Effect of climate change on insect pest management. In M. Coll and E. Wajnberg (Eds.), Environmental Pest Management: Challenges for Agronomists, Ecologists, Economists And Policymakers (1st ed.) (pp. 197–215). John Wiley & Sons Ltd.

Awmack, C., Woodcock, C., and Harrington, R. (1997). Climate change may increase vulnerability of aphids to natural enemies. Ecological Entomology, 22(3), 366–368. https://doi.org/10.1046/j.1365-2311.1997. 00069.x.

Bacon, S.J., Aebi, A., Calanca, P., and Bacher, S. (2014). Quarantine arthropod invasions in Europe: The role of climate, hosts and propagule pressure. Diversity and Distributions, 20(1), 84–94. https://doi.org /10.1111/ddi.12149.

Bale, J.S., Masters, G.J., Hodkinson, I.D., Awmack, C., Bezemer, T.M., Brown, V.K., Butterfield, J., Buse, A., Coulson, J.C., Farrar, J., Good, J.E.G., Harrington, R., Hartley, S., Jones, T.H., Lindroth, R.L., Press, M.C., Symrnioudis, I., Watt, A.D., and Whittaker, J.B. (2002). Herbivory in global climate change research: Direct effects of rising temperature on insect herbivores. Global Change Biology, 8(1), 1–16. https://doi.org/10.1046/j.1365-2486.2002.00451.x.

Banza, P., Evans, D.M., Medeiros, R., Macgregor, C.J., and Belo, A.D.F. (2021). Short-term positive effects of wildfire on diurnal insects and pollen transport in a Mediterranean ecosystem. Ecological Entomology, 46(6), 1353–1363. https://doi.org/10.1111/een.13082.

Barzman, M., Bàrberi, P., Birch, A.N.E., Boonekamp, P., Dachbrodt-Saaydeh, S., Graf, B., Hommel, B., Jensen, J.E., Kiss, J., Kudsk, P., Lamichhane, J.R., Messéan, A., Moonen, A., Ratnadass, A., Ricci, P., Sarah, J., and Sattin, M. (2015). Eight principles of integrated pest management. *Agronomy for Sustainable Development*, 35(4), 1199–1215. https://doi.org/10.1007/s13593-015-0327-9.

Bebber, D.P., Ramotowski, M.A.T., and Gurr, S.J. (2013). Crop pests and pathogens move polewards in a warming world. *Nature Climate Change*, 3(11), 985–988. https://doi.org/10.1038/nclimate1990.

Bernays, E.A. (1997). Feeding by lepidopteran larvae is dangerous. *Ecological Entomology*, 22(1), 121–123. https://doi.org/10.1046/j.1365-2311. 1997.00042.x.

Bezemer, T.M., Jones, T.H., and Knight, K.J. (1998). Long-term effects of elevated CO_2 and temperature on populations of the peach potato aphid Myzus persicae and its parasitoid Aphidius matricariae. *Oecologia*, 116(1–2), 128–135. https://doi.org/10.1007/s004420050571.

Bhoi, T.K., Samal, I., Majhi, P.K., Komal, J., Mahanta, D.K., Pradhan, A.K., Saini, V., Nikhil Raj, M., Ahmad, M.A., Behera, P.P., and Ashwini, M. (2022). Insight into aphid mediated potato virus Y transmission: A molecular to bioinformatics prospective. *Frontiers in Microbiology*, 13, 1001454. https://doi.org/10.3389/fmicb.2022.1001454.

Blanckenberg, M., Mlambo, M.C., Parker, D., and Reed, C. (2019). The negative impacts of fire on the resurrection ecology of invertebrates from temporary wetlands in cape flats sand fynbos in the Western Cape, South Africa. *Austral Ecology*, 44(7), 1225–1235. https://doi.org/10.11 11/aec.12800.

Boland, G.J., Melzer, M.S., Hopkin, A., Higgins, V., and Nassuth, A. (2004). Climate change and plant diseases in Ontario. *Canadian Journal of Plant Pathology*, 26(3), 335–350. https://doi.org/10.1080/07060 660409507151.

Boller, E.F., Avilla, J., Joerg, E., Malavolta, C., Wijnands, F.G., and Esbjerg, P. (2004). Integrated production: Principles and technical guidelines. https://www.iobc-wprs.org/ip_ipm/01_IOBC_Principles_an d_Tech_Guidelines_2004.pdf. Retrieved March 13, 2015. IOBC/WPRS.

Boudon-Padieu, E. (1932). Cicadelle vectrice de la flavescence dorée, Scaphoideus Titanus Ball. In S. Ravageurs de la Vigne and J. (Eds.), *Féret*, 2000 (pp. 110–120). Bordeaux, France.

Boudon-Padieu, E. (2005). Phytoplasmas associated to Grapevine yellows and potential vectors. *Boll. O.I.V.*, 78, 299–320.

Boudon-Padieu, E. (2003, September 12–17). The situation of grapevine yellows and current research directions: Distribution, diversity, vectors, diffusion and control. *Proceedings of the Extended Abstracts*, 14th Meeting of the ICVG, Locorotondo, Italy (pp. 47–53).

Boudon-Padieu, É. and Maixner, M. (2007, March 28–30). Potential effects of climate change on distribution and activity of insect vectors of grapevine pathogens. *Proceedings of the International and Multi-Disciplinary "Global Warming, Which Potential Impacts on the Vine-yards?,"* Beaune, France p. 23.

Brennan, K.E.C., Moir, M.L., and Wittkuhn, R.S. (2011). Fire refugia: The mechanism governing animal survivorship within a highly flammable plant. *Austral Ecology*, 36(2), 131–141. https://doi.org/10.1111/j.1442-9993.2010.02127.x.

Burgiel, S.W. and Muir, A.A. (2010). *Invasive Species, Climate Change and Ecosystem-Based Adaptation: Addressing Multiple Drivers of Global Change*. Global Invasive Species Program.

Campbell, J.W., Hanula, J.L., and Waldrop, T.A. (2007). Effects of pre-scribed fire and fire surrogates on floral visiting insects of the Blue Ridge province in North Carolina. *Biological Conservation*, 134(3), 393–404. https://doi.org/10.1016/j.biocon.2006.08.029.

Canto, T., Aranda, M.A., and Fereres, A. (2009). Climate change effects on physiology and population processes of hosts and vectors that influence the spread of hemipteran-borne plant viruses. *Global Change Biology*, 15(8), 1884–1894. https://doi.org/10.1111/j.1365-2486.2008.01820.x.

Cayton, H.L., Haddad, N.M., Gross, K., Diamond, S.E., and Ries, L. (2015). Do growing degree days predict phenology across butterfly species? *Ecology*, 96(6), 1473–1479. https://doi.org/10.1890/15-0131.1.

Community-based Distribution. Invasive alien species: The application of classical biological control for the management of established invasive alien species causing environmental impacts. *Proceedings of the Con-ference of the Parties to the Convention on Biological Diversity. Con-vention on Biological Diversity*, 14th Meeting, Sharm el-Sheikh, Egypt, November 17–29, 2018.

Chevin, L.M., Lande, R., and Mace, G.M. (2010). Adaptation, plasticity, and extinction in a changing environment: Towards a predictive theory. *PLOS Biology*, 8(4), e1000357. https://doi.org/10.1371/journal.pbio.1000357.

Cini, A., Anfora, G., Escudero-Colomar, L.A., Grassi, A., Santosuosso, U., Seljak, G., and Papini, A. (2014). Tracking the invasion of the alien fruit pest Drosophila suzukii in Europe. *Journal of Pest Science*, 87(4), 559–566. https://doi.org/10.1007/s10340-014-0617-z.

Collins, W., Colman, R., Haywood, J., Manning, M.R., and Mote, P. (2007). The physical science behind climate change. *Scientific American*, 297(2), 64–73. https://doi.org/10.1038/scientificamerican0807-64.

Cotrufo, M.F., Ineson, P., and Scott, A. (1998). Elevated CO2 reduces the nitrogen concentration of plant tissues. *Global Change Biology*, 4(1), 43–54. https://doi.org/10.1046/j.1365-2486.1998.00101.x.

Coulson, S.J., Hodkinson, I.D., Webb, N.R., Mikkola, K., Harrison, J.A., and Pedgley, D.E. (2002). Aerial colonization of high Arctic islands by invertebrates: The diamondback moth plutella xylostella (Lepidoptera: Yponomeutidae) as a potential indicator species. *Diversity and Distributions*, 8(6), 327–334. https://doi.org/10.1046/j.1472-4642.2002.00157.x.

Coviella, C.E. and Trumble, J.T. (1999). Effects of elevated atmospheric carbon dioxide on insect-plant interactions. *Conservation Biology*, 13(4), 700–712. https://doi.org/10.1046/j.1523-1739.1999.98267.x.

Dai, A. (2011). Drought under global warming: A review. *WIREs Climate Change*, 2(1), 45–65. https://doi.org/10.1002/wcc.81.

De Palma, A., Dennis, R.L.H., Brereton, T., Leather, S.R., and Oliver, T.H. (2017). Large reorganizations in butterfly communities during an extreme weather event. *Ecography*, 40(5), 577–585. https://doi.org/10.1111/ecog.02228

Deacon, C., Samways, M.J., and Pryke, J.S. (2019). Aquatic insects decline in abundance and occupy low-quality artificial habitats to survive hydrological droughts. *Freshwater Biology*, 64(9), 1643–1654. https://doi.org/10.1111/fwb.13360.

Dell, J.E., Salcido, D.M., Lumpkin, W., Richards, L.A., Pokswinski, S.M., Loudermilk, E.L., O'Brien, J.J., and Dyer, L.A. (2019). Interaction diversity maintains resiliency in a frequently disturbed ecosystem. *Frontiers in Ecology and Evolution*, 7, 145. https://doi.org/10.3389/fevo.2019.00145.

DeLucia, E.H., Casteel, C.L., Nabity, P.D., and O'Neill, B.F. (2008). Insects take a bigger bite out of plants in a warmer, higher carbon dioxide world. *Proceedings of the National Academy of Sciences of the United States of America*, 105(6), 1781–1782. https://doi.org/10.1073/pnas.0712056105.

Deutsch, C.A., Tewksbury, J.J., Huey, R.B., Sheldon, K.S., Ghalambor, C.K., Haak, D.C., and Martin, P.R. (2008). Impacts of climate warming on terrestrial ectotherms across latitude. *Proceedings of the National Academy of Sciences of the United States of America*, 105(18), 6668–6672. https://doi.org/10.1073/pnas.0709472105.

Deutsch, C.A., Tewksbury, J.J., Tigchelaar, M., Battisti, D.S., Merrill, S.C., Huey, R.B., and Naylor, R.L. (2018). Increase in crop losses to insect pests in a warming climate. *Science*, 361(6405), 916–919. https://doi.org/10.1126/science.aat3466.

Dukes, J.S.D.S., Pontius, J., Orwig, D., Garnas, J.R.G.R., Rodgers, V.L., Brazee, N., Cooke, B., Theoharides, K. A.T.A., Stange, E.E.S.E., Harrington, R., Ehrenfeld, J., Gurevitch, J., Lerdau, M., Stinson, K.,

Wick, R., and Ayres, M. (2009). Responses of insect pests, pathogens, and invasive plant species to climate change in the forests of northeastern North America: What can we predict? This article is one of a selection of papers from NE Forests 2100: A Synthesis of Climate Change Impacts on Forests of the Northeastern US and Eastern Canada. *Canadian Journal of Forest Research*, 39(2), 231–248. https://doi.org/10.1139/X08-171.

Ezcurra, E., Rapoport, E.H., and Marino, C.R. (1978). The geographical distribution of insect pests. *Journal of Biogeography*, 5(2), 149. https://doi.org/10.2307/3038169.

Falzoi, S., Lessio, F., Spanna, F., and Alma, A. (2014). Influence of temperature on the embryonic and post-embryonic development of Scaphoideus Titanus (Hemiptera: Cicadellidae), vector of grapevine Flavescence dorée. *International Journal of Pest Management*, 60(4), 246–257. https://doi.org/10.1080/09670874.2014.966170.

Fand, B.B., Kamble, A.L., and Kumar, M. (2012). Will climate change pose serious threat to crop pest management: A critical review. *International Journal of Science and Research*, 2, 1–14.

Food and Agriculture Organization. (2008). Climate related transboundary pests and diseases. http://www.fao.org/3/a-ai785e.pdf. Retrieved December 19, 2020.

Food and Agriculture Organization. (2019). Food and Agriculture Organization plant pests and diseases in the context of climate change and climate variability, food security and biodiversity risks. http://www.fao.org/3/nb088/nb088.pdf. Retrieved January 12, 2020.

Food and Agriculture Organization. How to practice integrated pest management. Available online. http://www.fao.org/agriculture/crops/thematic-sitemap/theme/compendium/tools-guidelines/how-to-ipm/en/. Retrieved February 15, 2021.

Fereres, A., Irwin, M.E., and Kampmeier, G.E. (2017). Aphid movement: Process and consequences. In H.F. van Emden and R. Harrington (Eds.), *Aphids as Crop Pests* (2nd ed.) (pp. 196–224). CABI Publishing.

Field, C.B., Barros, V.R., Dokken, D.J., Mach, K.J., Mastrandrea, M.D., Bilir, T.E., Chatterjee, M., Ebi, K.L., Estrada, Y.O., Genova, R.C. *et al.* (2014). IPCC Summary for policymakers. In *Climate Change: Impacts, Adaptation, and Vulnerability, Part, A.* (2014). Global and sectoral aspects; contribution of Working Group II to the fifth assessment report of the Intergovernmental Panel on Climate Change (pp. 1–32). Cambridge University Press.

Fleming, R.A. and Volney, W.J.A. (1995). Effects of climate change on insect defoliator population processes in Canada's boreal forest: Some plausible scenarios. *Water, Air, and Soil Pollution*, 82(1–2), 445–454. https://doi.org/10.1007/BF01182854.

Fox, R.J., Donelson, J.M., Schunter, C., Ravasi, T., and Gaitán-Espitia, J.D. (2019). Beyond buying time: The role of plasticity in phenotypic adaptation to rapid environmental change. *Philosophical Transactions of the Royal Society of London. Series B, Biological Sciences*, 374(1768), 20180174. https://doi.org/10.1098/rstb.2018.0174.

Fuhrer, J. (2003). Agroecosystem responses to combinations of elevated CO_2, ozone, and global climate change. *Agriculture, Ecosystems and Environment*, 97(1–3), 1–20. https://doi.org/10.1016/S0167-8809(03)00125-7.

Gely, C. (2021). *How Will Increased Drought Affect Insect Communities in Australian Tropical Rainforests?* [PhD Thesis] [Dissertation]. Griffith University.

Gely, C., Laurance, S.G.W., and Stork, N.E. (2020). How do herbivorous insects respond to drought stress in trees? *Biological Reviews of the Cambridge Philosophical Society*, 95(2), 434–448. https://doi.org/10.1111/brv.12571.

Gely, C., Laurance, S.G.W., and Stork, N.E. (2021). The effect of drought on wood-boring in trees and saplings in tropical rainforests. *Forest Ecology and Management*, 489, 119078. https://doi.org/10.1016/j.foreco.2021.119078.

Global Monitoring Division, and Earth System Research Laboratory. Trends in atmospheric carbon dioxide. http://www.esrl.noaa.gov/gmd/ccgg/trends/. Retrieved July 1, 2011.

Godfray, H.C.J., Beddington, J.R., Crute, I.R., Haddad, L., Lawrence, D., Muir, J.F., Pretty, J., Robinson, S., Thomas, S.M., and Toulmin, C. (2010). Food security: The challenge of feeding 9 billion people. *Science*, 327(5967), 812–818. https://doi.org/10.1126/science.1185383.

Gomez-Zavaglia, A., Mejuto, J.C., and Simal-Gandara, J. (2020). Mitigation of emerging implications of climate change on food production systems. *Food Research International*, 134, 109256. https://doi.org/10.1016/j.foodres.2020.109256.

Gregory, P.J., Johnson, S.N., Newton, A.C., and Ingram, J.S.I. (2009). Integrating pests and pathogens into the climate change/food security debate. *Journal of Experimental Botany*, 60(10), 2827–2838. https://doi.org/10.1093/jxb/erp080.

Gutbrodt, B., Mody, K., and Dorn, S. (2011). Drought changes plant chemistry and causes contrasting responses in lepidopteran herbivores. Oikos, 120(11), 1732–1740. https://doi.org/10.1111/j.1600-0706.2011.19558.x.

Gutierrez, A.P., D'Oultremont, T., Ellis, C.K., and Ponti, L. (2006). Climatic limits of pink bollworm in Arizona and California: Effects of climate warming. *Acta Oecologica*, 30(3), 353–364. https://doi.org/10.1016/j.actao.2006.06.003.

Gutierrez, A.P., Ponti, L., and Cossu, Q.A. (2009). Prospective comparative analysis of global warming effects on olive and olive fly. Climatic Change. (Bactrocera oleae (Gmelin)). https://doi.org/10.1007/s10584-008-9528-4 in Arizona–California and Italy. *Climatic Change*, 95(1–2), 195–217. https://doi.org/10.1007/s10584-008-9528-4.

Halsch, C.A., Shapiro, A.M., Fordyce, J.A., Nice, C.C., Thorne, J.H., Waetjen, D.P., and Forister, M.L. (2021). Insects and recent climate change. *Proceedings of the National Academy of Sciences of the United States of America*, 118(2), e2002543117. https://doi.org/10.1073/pnas.2002543117.

Hamilton, J.G., Dermody, O., Aldea, M., Zangerl, A.R., Rogers, A., Berenbaum, M.R., and DeLucia, E.H. (2005). Anthropogenic changes in tropospheric composition increase susceptibility of soybean to insect herbivory. *Environmental Entomology*, 34(2), 479–485. https://doi.org/10.1603/0046-225X-34.2.479.

Hance, T., Van Baaren, J., Vernon, P., and Boivin, G. (2007). Impact of extreme temperatures on parasitoids in a climate change perspective. *Annual Review of Entomology*, 52, 107–126. https://doi.org/10.1146/annurev.ento.52.110405.091333.

Harrington, R., Clark, S.J., Welham, S.J., Verrier, P.J., Denholm, C.H., Hullé, M., Maurice, D., Rounsevell, M. D., and Cocu, N. (2007). Environmental change and the phenology of European aphids. *Global Change Biology*, 13(8), 1550–1564. https://doi.org/10.1111/j.1365-2486.2007.01394.x.

Harris, R.M.B., Beaumont, L.J., Vance, T.R., Tozer, C.R., Remenyi, T.A., Perkins-Kirkpatrick, S.E., Mitchell, P.J., Nicotra, A.B., McGregor, S., Andrew, N.R., Letnic, M., Kearney, M.R., Wernberg, T., Hutley, L.B., Chambers, L.E., Fletcher, M.-S., Keatley, M.R., Woodward, C.A., Williamson, G., ... Bowman, D.M.J. S. (2018). Biological responses to the press and pulse of climate trends and extreme events. *Nature Climate Change*, 8(7), 579–587. https://doi.org/10.1038/s41558-018-0187-9.

Harvey, J.A., Heinen, R., Armbrecht, I., Basset, Y., Baxter-Gilbert, J.H., Bezemer, T.M., Böhm, M., Bommarco, R., Borges, P.A.V., Cardoso, P., Clausnitzer, V., Cornelisse, T., Crone, E.E., Dicke, M., Dijkstra, K.B., Dyer, L., Ellers, J., Fartmann, T., Forister, M.L., ... de Kroon, H. (2020). International scientists formulate a roadmap for insect conservation and recovery. *Nature Ecology and Evolution*, 4(2), 174–176. https://doi.org/10.1038/s41559-019-1079-8.

Heeb, L., Jenner, E., and Cock, M.J.W. (2019). Climate-smart pest management: Building resilience of farms and landscapes to changing pest threats. *Journal of Pest Science*, 92(3), 951–969. https://doi.org/10.1 007/s10340-019-01083-y.

Hellmann, J.J., Byers, J.E., Bierwagen, B.G., and Dukes, J.S. (2008). Five potential consequences of climate change for invasive species. *Conservation Biology*, 22(3), 534–543. https://doi.org/10.1111/j.1523-1739.2 008.00951.x.

Heuskin, S., Verheggen, F.J., Haubruge, E., Wathelet, J.P., and Lognay, G. (2011). The use of semiochemical slow-release devices in integrated pest management strategies. *Biotechnology, Agronomy and Society and Environment*, 15, 459–470.

Hill, D.S. (1987). *Agricultural Insect Pests of Temperate Regions and Their Control*. Cambridge University Press.

Hill, M.P., Bertelsmeier, C., Clusella-Trullas, S., Garnas, J., Robertson, M.P., and Terblanche, J.S. (2016). Predicted decrease in global climate suitability masks regional complexity of invasive fruit fly species response to climate change. *Biological Invasions*, 18(4), 1105–1119. https://doi.org/10.1007/s10530-016-1078-5.

Howden, S.M., Soussana, J.F., Tubiello, F.N., Chhetri, N., Dunlop, M., and Meinke, H. (2007). Adapting agriculture to climate change. *Proceedings of the National Academy of Sciences of the United States of America*, 104(50), 19691–19696. https://doi.org/10.1073/pnas.0701890104.

Hull, R. (2014). *Plant Virology*. Elsevier/Academic Press.

Hullé, M., Coeur d'Acier, A.C., Bankhead-Dronnet, S., and Harrington, R. (2010). Aphids in the face of global changes. *Comptes Rendus Biologies*, 333(6–7), 497–503. https://doi.org/10.1016/j.crvi.2010.03.005.

Irwin, M.E., Ruesink, W.G., Isard, S.A., and Kampmeier, G.E. (2000). Mitigating epidemics caused by non-persistently transmitted aphidborne viruses: The role of the pliant environment. *Virus Research*, 71(1–2), 185–211. https://doi.org/10.1016/s0168-1702(00)00198-2.

Jain, P., Castellanos-Acuna, D., Coogan, S.C.P., Abatzoglou, J.T., and Flannigan, M.D. (2022). Observed increases in extreme fire weather driven by atmospheric humidity and temperature. *Nature Climate Change*, 12(1), 63–70. https://doi.org/10.1038/s41558-021-01224-1.

Janzen, D.H. and Hallwachs, W. (2021). To us insectometers, it is clear that insect decline in our Costa Rican tropics is real, so let's be kind to the survivors. *Proceedings of the National Academy of Sciences of the United States of America*, 118(2), e2002546117. https://doi.org/ 10.1073/pnas.2002546117.

Jarošová, J., Żelazny, W.R., and Kundu, J.K. (2019). Patterns and predictions of barley yellow dwarf virus vector migrations in Central Europe. *Plant Disease*, 103(8), 2057–2064. https://doi.org/10.1094/PDIS-11-18-1999-RE.

Jinbo, U., Kato, T., and Ito, M. (2011). Current progress in DNA barcoding and future implications for entomology. *Entomological Science*, 14(2), 107–124. https://doi.org/10.1111/j.1479-8298.2011.00449.x.

Joern, A. and Laws, A.N. (2013). Ecological mechanisms underlying arthropod species diversity in grasslands. *Annual Review of Entomology*, 58, 19–36. https://doi.org/10.1146/annurev-ento-120811-153540.

Johnson, S.N., Anderson, E.A., Dawson, G., and Griffiths, D.W. (2008). Varietal susceptibility of potatoes to wireworm herbivory. *Agricultural and Forest Entomology*, 10(2), 167–174. https://doi.org/10.1111/j.1461-9563.2008.00372.x.

Jooste, M.L., Samways, M.J., and Deacon, C. (2020). Fluctuating pond water levels and aquatic insect persistence in a drought-prone Mediterranean-type climate. *Hydrobiologia*, 847(5), 1315–1326. https://doi.org/10.1007/s10750-020-04186-1.

Joyce, L.A., Briske, D.D., Brown, J.R., Polley, H.W., McCarl, B.A., and Bailey, D.W. (2013). Climate change and North American rangelands: Assessment of mitigation and adaptation strategies. *Rangeland Ecology and Management*, 66(5), 512–528. https://doi.org/10.2111/REM-D-12-00142.1.

Kiritani, K., and Yamamura, K. (2003). Exotic insects and their pathways for invasion. In G. M. Ruiz and J. T. Carlton (Eds.), *Invasive Species: Vectors and Management Strategies* (pp. 44–67). Island Press.

Kocmánková, E., Trnka, M., Juroch, J., Dubrovský, M., Semerádová, D., Možný, M., Žalud, Z., Pokorný, R., and Lebeda, A. (2010). Impact of climate change on the occurrence and activity of harmful organisms. *Plant Protection Science*, 45, S48–S52.

Koltz, A.M., Burkle, L.A., Pressler, Y., Dell, J.E., Vidal, M.C., Richards, L.A., and Murphy, S.M. (2018). Global change and the importance of fire for the ecology and evolution of insects. *Current Opinion in Insect Science*, 29, 110–116. https://doi.org/10.1016/j.cois.2018.07.015.

Kral, K.C., Limb, R.F., Harmon, J.P., and Hovick, T.J. (2017). Arthropods and fire: Previous research shaping future conservation. *Rangeland Ecology and Management*, 70(5), 589–598. https://doi.org/10.1016/j.rama.2017.03.006.

Lamichhane, J.R., Barzman, M., Booij, K., Boonekamp, P., Desneux, N., Huber, L., Kudsk, P., Langrell, S.R. H., Ratnadass, A., Ricci, P., Sarah, J., and Messéan, A. (2015). Robust cropping systems to tackle pests under climate change. A review. *Agronomy for Sustainable Development*, 35(2), 443–459. https://doi.org/10.1007/s13593-014-0275-9.

Laštůvka, Z. (2010). Climate change and its possible influence on the occurrence and importance of insect pests. *Plant Protection Science*, 45, S53–S62.

Lehmann, P., Ammunét, T., Barton, M., Battisti, A., Eigenbrode, S.D., Jepsen, J.U., Kalinkat, G., Neuvonen, S., Niemelä, P., Terblanche, J.S., Økland, B., and Björkman, C. (2020). Complex responses of global insect pests to climate warming. *Frontiers in Ecology and the Environment*, 18(3), 141–150. https://doi.org/10.1002/fee.2160.

Lincoln, D.E. (1993). The influence of plant carbon dioxide and nutrient supply on susceptibility to insect herbivores. *Vegetatio*, 104–105(1), 273–280. https://doi.org/10.1007/BF00048158.

Lincoln, D.E., Sionit, N., and Strain, B.R. (1984). Growth and feeding response of Pseudoplusia includens (Lepidoptera: Noctuidae) to host plants grown in controlled carbon dioxide atmospheres. *Environmental Entomology*, 13(6), 1527–1530. https://doi.org/10.1093/ee/13.6.1527.

Lindroth, R.L., Kinney, K.K., and Platz, C.L. (1993). Responses of diciduous trees to elevated atmospheric CO2: Productivity, phytochemistry, and insect performance. *Ecology*, 74(3), 763–777. https://doi.org/10.2307/1940804.

Lockwood, J.L., Cassey, P., and Blackburn, T.M. (2005). The role of propagule pressure in explaining species invasions. *Trends in Ecology and Evolution*, 20(5), 223–228. https://doi.org/10.1016/j.tree.2005.02.004.

Lopez-Vaamonde, C., Agassiz, D., Augustin, S., De Prins, J., De Prins, W., De Prins, W., Gomboc, S., Ivinskis, P., Karsholt, O., Koutroumpas, A., Koutroumpa, F., Laštůvka, Z., Marabuto, E., Olivella, E., Przybylowicz, L., Roques, A., Ryrholm, N., Sefrova, H., ... Lees, D. (2010). Lepidoptera. *BioRisk*, 4, 603–668. https://doi.org/10.3897/biorisk.4.50.

Mafokoane, L.D., Zimmermann, H.G., and Hill, M.P. (2007). Development of Cactoblastis cactorum (Berg) (Lepidoptera: Pyralidae) on six North American Opuntia species. *African Entomology*, 15(2), 295–299. https://doi.org/10.4001/1021-3589-15.2.295.

Mahanta, D.K., Jangra, S., Priti, Ghosh, A., Sharma, P.K., Iquebal, M.A., Jaiswal, S., Baranwal, V.K., Kalia, V. K., and Chander, S. (2022). Groundnut bud necrosis virus modulates the expression of innate immune, endocytosis, and cuticle development-associated genes to circulate and propagate in its vector, Thrips palmi. *Frontiers in Microbiology*, 13, 773238. https://doi.org/10.3389/fmicb.2022.773238.

Mahanta, D.K., Komal, J., Samal, I., Bhoi, T.K., Dubey, V.K., Pradhan, K., Nekkanti, A., Gouda, M.N.R., Saini, V., Negi, N., Bhateja, S., Jat, H. K., Jeengar, D., ... Jeengar, D., ... and Jeengar11. Nutritional aspects and dietary benefits of "Silkworms": Current scenario and future outlook. *Frontiers in Nutrition*, 10, 44. https://doi.org/10.3389/fnut.2023 .1121508.

Mahlman, J.D. (1997). Uncertainties in projections of human-caused climate warming. *Science*, 278(5342), 1416–1417. https://doi.org/10.1126/science.278.5342.1416.

Martín-Vertedor, D., Ferrero-García, J.J., and Torres-Vila, L.M. (2010). Global warming affects phenology and voltinism of Lobesia botrana in Spain. *Agricultural and Forest Entomology*, 12(2), 169–176. https://doi.org/10.1111/j.1461-9563.2009.00465.x.

Masters, G. and Norgrove, L. (2010). Climate change and invasive alien species. *CABI Working Paper*, 1, p. 30.

Menéndez, R. (2007). How are insects responding to global warming? *Tijdschrift voor Entomologie*, 150, 355.

Menéndez, R., González-Megías, A., Collingham, Y., Fox, R., Roy, D.B., Ohlemüller, R., and Thomas, C.D. (2007). Direct and indirect effects of climate and habitat factors on butterfly diversity. *Ecology*, 88(3), 605–611. https://doi.org/10.1890/06-0539.

Meynard, C.N., Migeon, A., and Navajas, M. (2013). Uncertainties in predicting species distributions under climate change: A case study using Tetranychus evansi (Acari: Tetranychidae), a widespread agricultural pest. *PLOS ONE*, 8(6), e66445. https://doi.org/10.1371/journal.pone.0066445.

Mirutenko, V., Janský, V., and Margitay, V. (2018). First records of Scaphoideus Titanus (Hemiptera, Cicadellidae) in Ukraine. *EPPO Bulletin*, 48(1), 167–168. https://doi.org/10.1111/epp.12460.

Nault, L.R. (1997). Arthropod transmission of plant viruses: A new synthesis. *Annals of the Entomological Society of America*, 90(5), 521–541. https://doi.org/10.1093/aesa/90.5.521.

Nihal, R. (2020). Global Climate change and its impact on integrated pest management. *Agro Econ. International Journal*, 7, 133–137.

Nimmo, D.G., Carthey, A.J.R., Jolly, C.J., and Blumstein, D.T. (2021). Welcome to the Pyrocene: Animal survival in the age of megafire. *Global Change Biology*, 27(22), 5684–5693. https://doi.org/10.1111/gcb.15834.

Pareek, A., Meena, B.M., Sharma, S., Tetarwal, M.L., Kalyan, R.K., and Meena, B.L. (2017). Impact of climate change on insect pests and their management strategies. In P.S. Kumar, M. Kanwat, P.D. Meena,

V. Kumar and R.A. Alone (Eds.), *Climate Change and Sustainable Agriculture* (pp. 253–286). New India Publishing Agency.

Parmesan, C., Ryrholm, N., Stefanescu, C., Hill, J.K., Thomas, C.D., Descimon, H., Huntley, B., Kaila, L., Kullberg, J., Tammaru, T., Tennent, W.J., Thomas, J.A., and Warren, M. (1999). Poleward shifts in geographical ranges of butterfly species associated with regional warming. *Nature*, 399(6736), 579–583. https://doi.org/10.1038/21181.

Parry, H.R., Macfadyen, S., and Kriticos, D.J. (2012). The geographical distribution of Yellow dwarf viruses and their aphid vectors in Australian grasslands and wheat. *Australasian Plant Pathology*, 41(4), 375–387. https://doi.org/10.1007/s13313-012-0133-7.

Parry, M. (1990). The potential impact on agriculture of the greenhouse effect. *Land Use Policy*, 7(2), 109–123. https://doi.org/10.1016/0264-8377(90)90003-H.

Pathak, H., Aggarwal, P.K., and Singh, S.D. (2012). *Climate Change Impact, Adaptation and Mitigation in Agriculture: Methodology for Assessment and Applications.* Indian Agricultural Research Institute.

Pathania, M., Verma, A., Singh, M., Arora, P.K., and Kaur, N. (2020). Influence of abiotic factors on the infestation dynamics of whitefly, Bemisia tabaci (Gennadius 1889) in cotton and its management strategies in North-Western India. *International Journal of Tropical Insect Science*, 40(4), 969–981. https://doi.org/10.1007/s42690-020-00155-2.

Patterson, D.T., Westbrook, J.K., Joyce, R.J.V., Lingren, P.D., and Rogasik, J. (1999). Weeds, insects, and diseases. *Climatic Change*, 43(4), 711–727. https://doi.org/10.1023/A:1005549400875.

Perrings, C., Burgiel, S., Lonsdale, M., Mooney, H., and Williamson, M. (2010). International cooperation in the solution to trade-related invasive species risks. *Annals of the New York Academy of Sciences*, 1195, 198–212. https://doi.org/10.1111/j.1749-6632.2010.05453.x.

Pollard, E. and Yates, T.J. (1993). *Monitoring Butterflies for Ecology and Conservation: The British Butterfly Monitoring Scheme.* Springer.

Porter, J.H., Parry, M.L., and Carter, T.R. (1991). The potential effects of climatic change on agricultural insect pests. *Agricultural and Forest Meteorology*, 57(1–3), 221–240. https://doi.org/10.1016/0168-1923(91)90088-8.

Poyet, M., Le Roux, V., Gibert, P., Meirland, A., Prévost, G., Eslin, P., and Chabrerie, O. (2015). The wide potential trophic niche of the Asiatic fruit fly Drosophila suzukii: The key of its invasion success in temperate Europe? *PLOS ONE*, 10(11), e0142785. https://doi.org/10.1371/journal.pone.0142785.

Prakash, A., Rao, J., Mukherjee, A.K., Berliner, J., Pokhare, S.S., Adak, T., Munda, S., and Shashank, P.R. (2014). *Climate Change: Impact on*

Crop Pests. Applied Zoologists Research Association, Central Rice Research Institute.

Prentice, I.C., Farquhar, G.D., Fasham, M.J.R., Goulden, M.L., Heimann, M., Jaramillo, V.J., Kheshgi, H.S., Le Quéré, C., Scholes, R.J., and Wallace, D.W.R. (2001). The scientific basis. The carbon cycle and atmospheric carbon dioxide. J.T. Houghton *et al.* (Eds.). In *Climate Change*. Cambridge University Press, 185–237.

Pryke, J.S. and Samways, M.J. (2012a). Importance of using many taxa and having adequate controls for monitoring impacts of fire for arthropod conservation. *Journal of Insect Conservation*, 16(2), 177–185. https://doi.org/10.1007/s10841-011-9404-9.

Pryke, J.S. and Samways, M.J. (2012b). Differential resilience of invertebrates to fire. *Austral Ecology*, 37(4), 460–469. https://doi.org/10.1111/j.1442-9993.2011.02307.x.

Ramos, R.S., Kumar, L., Shabani, F., da Silva, R.S., de Araújo, T.A., and Picanço, M.C. (2019). Climate model for seasonal variation in Bemisia tabaci using CLIMEX in tomato crops. *International Journal of Biometeorology*, 63(3), 281–291. https://doi.org/10.1007/s00484-018-01661-2.

Raza, M.M., Khan, M.A., Arshad, M., Sagheer, M., Sattar, Z., Shafi, J., Haq, Eu, Ali, A., Aslam, U., Mushtaq, A., Ishfaq, I., Sabir, Z., and Sattar, A. (2015). Impact of global warming on insects. *Archives of Phytopathology and Plant Protection*, 48(1), 84–94. https://doi.org/10.1080/03235408.2014.882132.

Rering, C.C., Franco, J.G., Yeater, K.M., and Mallinger, R.E. (2020). Drought stress alters floral volatiles and reduces floral rewards, pollinator activity, and seed set in a global plant. *Ecosphere*, 11(9), e03254. https://doi.org/10.1002/ecs2.3254.

Ricciardi, A. (2013). Invasive species. In R. Leemans (Ed.), *Ecological Systems* (1st. ed) (pp. 161–178). Springer.

Robert, Y., Woodford, J.A., and Ducray-Bourdin, D.G. (2000). Some epidemiological approaches to the control of aphid-borne virus diseases in seed potato crops in northern Europe. *Virus Research*, 71(1–2), 33–47. https://doi.org/10.1016/s0168-1702(00)00186-6.

Roff, D. (1980). Optimizing development time in a seasonal environment: The "ups and downs" of clinal variation. *Oecologia*, 45(2), 202–208. https://doi.org/10.1007/BF00346461.

Rogelj, J.D., Shindell, K., Jiang, S., Fifita, P., Forster, V., Ginzburg, C., Handa, H., Kheshgi, S., Kobayashi, E., Kriegler, E. *et al.* (2018). Mitigation pathways compatible with 1.5°C in the context of sustainable development. In V.P. Masson-Delmotte *et al.* (Eds.), *Global warming of, 1(5)° C.* An IPCC Special Report on the Impacts of Global Warming of 1.5°C above Pre-Industrial Levels and Related Global Greenhouse

Gas Emission Pathways, in the Context of Strengthening the Global Response to the Threat of Climate Change, Sustainable Development, and Efforts to Eradicate Poverty. World Meteorological Organization.

Rohde, K., Hau, Y., Kranz, N., Weinberger, J., Elle, O., and Hochkirch, A. (2017). Climatic effects on population declines of a rare wetland species and the role of spatial and temporal isolation as barriers to hybridization. *Functional Ecology*, 31(6), 1262–1274. https://doi.org/10.1111/1365-2435.12834.

Roos, J., Hopkins, R., Kvarnheden, A., and Dixelius, C. (2011). The impact of global warming on plant diseases and insect vectors in Sweden. *European Journal of Plant Pathology*, 129(1), 9–19. https://doi.org/10.1007/s10658-010-9692-z.

Rosenzweig, C. (1989). Global climate change: Predictions and observations. *American Journal of Agricultural Economics*, 71(5), 1265–1271. https://doi.org/10.2307/1243119.

Rosenzweig, C., Major, D.C., Demong, K., Stanton, C., Horton, R., and Stults, M. (2007). Managing climate change risks in New York City's water system: Assessment and adaptation planning. *Mitigation and Adaptation Strategies for Global Change*, 12(8), 1391–1409. https://doi.org/10.1007/s11027-006-9070-5.

Rota-Stabelli, O., Blaxter, M., and Anfora, G. (2013). Drosophila suzukii. *Current Biology*, 23(1), R8–R9. https://doi.org/10.1016/j.cub.2012.11.021.

Roth, S.K. and Lindroth, R.L. (1995). Elevated atmospheric CO_2: Effects on phytochemistry, insect performance and insect-parasitoid interactions. *Global Change Biology*, 1(3), 173–182. https://doi.org/10.1111/j.1365-2486.1995.tb00019.x.

Roy, D.B. and Sparks, T.H. (2000). Phenology of British butterflies and climate change. *Global Change Biology*, 6(4), 407–416. https://doi.org/10.1046/j.1365-2486.2000.00322.x.

Sastry, K.S. and Zitter, T.A. (2014). *Plant Virus and Viroid Diseases in the Tropics, 2: Epidemiology and Management* (1st edn.). Springer.

Saunders, M.E., Barton, P.S., Bickerstaff, J.R.M., Frost, L., Latty, T., Lessard, B.D., Lowe, E.C., Rodriguez, J., White, T.E., and Umbers, K.D.L. (2021). Limited understanding of bushfire impacts on Australian invertebrates. *Insect Conservation and Diversity*, 14(3), 285–293. https://doi.org/10.1111/icad.12493.

Sconiers, W. B., and Eubanks, M. D. (2017). Not all droughts are created equal? The effects of stress severity on insect herbivore abundance. *Arthropod-Plant Interactions*, 11(1), 45–60. https://doi.org/10.1007/s11829-016-9464-6.

Sgrò, C.M., Terblanche, J.S., and Hoffmann, A.A. (2016). What can plasticity contribute to insect responses to climate change? *Annual Review of Entomology*, 61, 433–451. https://doi.org/10.1146/annurev-ento-01 0715-023859.

Sharma, H.C. (2014). Climate change effects on insects: Implications for crop protection and food security. *Journal of Crop Improvement*, 28(2), 229–259. https://doi.org/10.1080/15427528.2014.881205.

Sharma, H.C., Dhillon, M.K., Kibuka, J., and Mukuru, S.Z. (2005). Plant defense responses to sorghum spotted stem borer, Chilo partellus under irrigated and drought conditions. *Int. Sorghum millets Newsletter 2005*, 46, 49–52.

Shine, C., Williams, N., and Gündling, L. (2000). A guide to designing legal and institutional frameworks on alien invasive species, environmental policy and law, *paper no. 40*. International Union for Conservation of Nature and Natural Resources.

Shrestha, S. (2019). Effects of climate change in agricultural insect pest. *Acta Scientific Agriculture*, 3(12), 74–80. https://doi.org/10.31080/ASAG.2019.03.0727.

Simberloff, D. (2009). The role of propagule pressure in biological invasions. *Annual Review of Ecology, Evolution, and Systematics*, 40(1), 81–102. https://doi.org/10.1146/annurev.ecolsys.110308.120304.

Skendžić, S., Zovko, M., Živković, I.P., Lešić, V., and Lemić, D. (2021). The impact of climate change on agricultural insect pests. *Insects*, 12, 440. https://doi.org/10.3390/insects12050440

Snell-Rood, E.C., Kobiela, M.E., Sikkink, K.L., and Shephard, A.M. (2018). Mechanisms of plastic rescue in novel environments. *Annual Review of Ecology, Evolution, and Systematics*, 49(1), 331–354. https://doi.org/10.1146/annurev-ecolsys-110617-062622.

Sseruwagi, P., Sserubombwe, W.S., Legg, J.P., Ndunguru, J., and Thresh, J.M. (2004). Methods of surveying the incidence and severity of cassava mosaic disease and whitefly vector populations on cassava in Africa: A review. *Virus Research*, 100(1), 129–142. https://doi.org/10.1016/j.virusres.2003.12.021.

Staley, J.T., Hodgson, C.J., Mortimer, S.R., Morecroft, M.D., Masters, G.J., Brown, V.K., and Taylor, M.E. (2007). Effects of summer rainfall manipulations on the abundance and vertical distribution of herbivorous soil macro-invertebrates. *European Journal of Soil Biology*, 43(3), 189–198. https://doi.org/10.1016/j.ejsobi.2007.02.010.

Stefanescu, C., Peñuelas, J., and Filella, I. (2003). Effects of climatic change on the phenology of butterflies in the northwest Mediterranean Basin. *Global Change Biology*, 9(10), 1494–1506. https://doi.org/10.1046/j.1365-2486.2003.00682.x.

Stiling, P., and Cornelissen, T. (2007), 13. How does elevated carbon dioxide, CO2 affect plant–herbivore interactions? A field experiment and meta-analysis of CO2-mediated changes on plant chemistry and herbivore performance. *Global Change Biology*, 13(9), 1823–1842.

Streck, N.A. (2005). Climate change and agroecosystems: The effect of elevated atmospheric CO2 and temperature on crop growth, development, and yield. *Ciência Rural. Rural*, 35(3), 730–740. https://doi.org/10.1590/S0103-84782005000300041.

Sun, Y. and Ge, F. (2011). How do aphids respond to elevated CO2? *Journal of Asia-Pacific Entomology*, 14(2), 217–220. https://doi.org/10.1016/j.aspen.2010.08.001.

Sutherst, R.W., Constable, F., Finlay, K.J., Harrington, R., Luck, J., and Zalucki, M.P. (2011). Adapting to crop pest and pathogen risks under a changing climate. *WIREs Climate Change*, 2(2), 220–237. https://doi.org/10.1002/wcc.102.

Thomson, L.J., Macfadyen, S., and Hoffmann, A.A. (2010). Predicting the effects of climate change on natural enemies of agricultural pests. *Biological Control*, 52(3), 296–306. https://doi.org/10.1016/j.biocontrol.2009.01.022.

Tilman, D., Balzer, C., Hill, J., and Befort, B.L. (2011). Global food demand and the sustainable intensification of agriculture. *Proceedings of the National Academy of Sciences of the United States of America*, 108(50), 20260–20264. https://doi.org/10.1073/pnas.1116437108.

Tobin, P.C., Nagarkatti, S., Loeb, G., and Saunders, M.C. (2008). Historical and projected interactions between climate change and insect voltinism in a multivoltine species. *Global Change Biology*, 14(5), 951–957. https://doi.org/10.1111/j.1365-2486.2008.01561.x.

Tobin, P.C., Parry, D., and Aukema, B.H. (2014). The influence of climate change on insect invasions in temperate forest ecosystems. *Forest Sciences*, 81, 267–293.

Trebicki, P. (2020). Climate change and plant virus epidemiology. *Virus Research*, 286, 198059. https://doi.org/10.1016/j.virusres.2020.198059.

Trębicki, P., Vandegeer, R.K., Bosque-Pérez, N.A., Powell, K.S., Dader, B., Freeman, A.J., Yen, A.L., Fitzgerald, G.J., and Luck, J.E. (2016). Virus infection mediates the effects of elevated CO_2 on plants and vectors. *Scientific Reports*, 6, 22785. https://doi.org/10.1038/srep22785.

Trumble, J.T. and Butler, C.D. (2009). Climate change will exacerbate California's insect pest problems. *California Agriculture*, 63(2), 73–78. https://doi.org/10.3733/ca.v063n02p73.

Vadez, V., Berger, J.D., Warkentin, T., Asseng, S., Ratnakumar, P., Rao, K.P.C., Gaur, P.M., Munier-Jolain, N., Larmure, A., Voisin, A.-S., Sharma, H.C., Pande, S., Sharma, M., Krishnamurthy, L., and Zaman,

M.A. (2012). Adaptation of grain legumes to climate change: A review. *Agronomy for Sustainable Development*, 32(1), 31–44. https://doi.org/10.1007/s13593-011-0020-6.

Vander Zanden, M.J. (2005). The success of animal invaders. *Proceedings of the National Academy of Sciences of the United States of America*, 102(20), 7055–7056. https://doi.org/10.1073/pnas.0502549102.

Vanhanen, H. Invasive insects in Europe — The role of climate change and global trade [Diss]. *Dissertationes Forestales*, 2008(57). https://doi.org/10.14214/df.57.

Vermeij, G.J. (1996). An agenda for invasion biology. *Biological Conservation*, 78(1–2), 3–9. https://doi.org/10.1016/0006-3207(96)00013-4.

Wagner, D.L. (2020). Insect declines in the Anthropocene. *Annual Review of Entomology*, 65, 457–480. https://doi.org/10.1146/annurev-ento-011019-025151.

Wagner, D.L. and Balowitz, R. (2021). The 2020–2021 megadrought and Southeastern Arizona's butterflies. *News of the Lepidopterists' Society*, 63, 202–205.

Walther, G.R., Roques, A., Hulme, P.E., Sykes, M.T., Pyšek, P., Kühn, I., Zobel, M., Bacher, S., Botta-Dukát, Z., Bugmann, H., Czúcz, B., Dauber, J., Hickler, T., Jarošík, V., Kenis, M., Klotz, S., Minchin, D., Moora, M., Nentwig, W., ... Settele, J. (2009). Alien species in a warmer world: Risks and opportunities. *Trends in Ecology and Evolution*, 24(12), 686–693. https://doi.org/10.1016/j.tree.2009.06.008.

Ward, N.L., and Masters, G.J. (2007). Linking climate change and species invasion: An illustration using insect herbivores. *Global Change Biology*, 13(8), 1605–1615. https://doi.org/10.1111/j.1365-2486.2007.01399.x.

Wenda-Piesik, A., Piesik, D., Nowak, A., and Wawrzyniak, M. (2016). Tribolium confusum responses to blends of cereal kernels and plant volatiles. *Journal of Applied Entomology*, 140(7), 558–563. https://doi.org/10.1111/jen.12284.

Williams, A.P., Cook, B.I., and Smerdon, J.E. (2022). Rapid intensification of the emerging southwestern North American megadrought in 2020–2021. *Nature Climate Change*, 12(3), 232–234. https://doi.org/10.1038/s41558-022-01290-z.

Yamamura, K. and Kiritani, K. (1998). A simple method to estimate the potential increase in the number of generations under global warming in temperate zones. *Applied Entomology and Zoology*, 33(2), 289–298. https://doi.org/10.1303/aez.33.289.

Yamamura, K., Yokozawa, M., Nishimori, M., Ueda, Y., and Yokosuka, T. (2006). How to analyze long-term insect population dynamics under climate change: 50-year data of three insect pests in paddy fields.

Population Ecology, 48(1), 31–48. https://doi.org/10.1007/s10144-005-0239-7.

Yihdego, Y., Salem, H.S., and Muhammed, H.H. (2019). Agricultural pest management policies during drought: Case studies in Australia and the State of Palestine. *Natural Hazards Review*, 20(1), 05018010. https://doi.org/10.1061/(ASCE)NH.1527-6996.0000312.

Yoro, K.O. and Daramola, M.O. (2020) Chapter 1. CO_2 emission sources, greenhouse gases, and the global warming effect. In M.R. Rahimpour, M. Farsi and M.A. Makarem (Eds.), *Advances in Carbon Capture* (1st edn.) (pp. 3–28). Woodhead Publishing.

Zeyen, R.J., Stromberg, E.L., and Kuehnast, E.L. (1987). Long-range aphid transport hypothesis for maize dwarf mosaic virus: History and distribution in Minnesota, USA. *Annals of Applied Biology*, 111(2), 325–336. https://doi.org/10.1111/j.1744-7348.1987.tb01459.x.

Ziska, L.H., Blumenthal, D.M., Runion, G.B., Hunt, E.R., Jr., and Diaz-Soltero, H. (2011). Invasive species and climate change: An agronomic perspective. *Climatic Change*, 105(1–2), 13–42. https://doi.org/10.1007/s10584-010-9879-5.

Zvereva, E.L. and Kozlov, M.V. (2006). Consequences of simultaneous elevation of carbon dioxide and temperature for plant–herbivore interactions: A meta analysis. *Global Change Biology*, 12(1), 27–41. https://doi.org/10.1111/j.1365-2486.2005.01086.x.

Chapter 6

Improvising Farmers' Income through Climate Resilient Potato Varieties for Cereals-Based System of Indo-Gangetic Plains

Pooja Pandey

*Agronomist, International Potato Center (CIP),
Region SWCA, New Delhi, Pusa Campus, India*
P.pandey@cgiar.org

Abstract

Potato is an important cash crop for farmers in the sub-tropical north-eastern and Indo-Gangetic plains of India. These regions have fertile soil with a suitable climate which produces good-quality potatoes. The crop is grown in the winter season. However, in the last few years these regions have suffered from climate change's impact. The unseasonal heavy rain during the planting or harvesting, short winters, high temperatures, and diseases significantly reduced the quality and yield of the crop, consequently hitting farmers' income. This impact can be reduced by adopting smart cultural practices like intercropping, mulching, micro-irrigation, and the use of climate-resilient short-duration varieties that can be fitted into cereal-based cropping systems.

Keywords: Climate change, Crop diversification, Germplasms, Improved varieties.

Introduction

Potato is an important food and cash crop in India. It is considered a
wholesome food as it provides carbohydrates, proteins, and vitamins
which significantly mitigate malnutrition. India is the second largest
producer of potatoes and produces 51.31 million tons of potatoes
from a 2.42 million hectare area (NHB, 2018). Nearly 80% of the
potatoes in India are grown in the plains during the winter season
under a short photoperiod. The remaining area is cultivated in summer under a long photoperiod in hills and plateaus. Farmers having
small to marginal land holdings particularly in Bihar, Assam, and
West Bengal grow potatoes both for food and income generation to
meet livelihood expenses. The varietal requirements of these different
regions or states vary due to their varying agroecological conditions.
In Bihar, potato is grown in all districts, however, a few districts like
Nalanda, Patna, Saran, Samastipur, Vaishali, Gopalganj, East and
West Champaran, Muzaffarpur, and Gaya account for 80% of the
total potato area in the state.

Potato area and potato productivity have increased significantly
in the last two decades. In spite of increased production, the potato
crop is facing challenges due to intensive cultivation, pressure on agricultural land, and increased pest and disease populations because of
climate change making potato cultivation expensive. Global warming is affecting the shifting and uneven pattern of rainfall, thereby
increasing incidents of extreme weather events like floods, droughts,
and frosting, which adversely impact potato production. The average
global temperature will increase between 1.4°C and 5.8°C over the
period 1990–2100. According to IPCC report, crop productivity is
expected to increase slightly at mid to high latitudes as local mean
temperature increases by 1–3°C, depending upon the crop, and then
decrease beyond that in some regions. At lower latitudes, especially
in seasonally dry and tropical regions, crop productivity is projected
to decrease even for small local temperature increases (1–2°C), leading to a growing risk of hunger. In potato cultivations, the night
temperature has a crucial influence on both tuber initiation and
starch deposition in tubers. The ideal temperature range is between
15°C and 18°C. Overnight temperatures above 22°C severely hamper tuber initiation and development. Introduction of heat-tolerant

potato varieties would increase potential yields by more than 5% in most potato production zones.

Diversification in Cropping Pattern and Potato Niche

Potato is an important cash crop for farmers in the sub-tropical northeastern and Indo-Gangetic plains of India. Potato crop is cultivated during the winter season between November/December to February/March. Among the Indian states, West Bengal, Bihar, and Assam produce a significant amount of rice. The *kharif* rice and wheat crops both remain about 140 days in the field. Even, the summer boro rice grown by traditional practices remains 100 days in the field. Whereas, potato is a 90-day crop and a good option for an intensive cropping system. It gives higher returns than most of the other crops.

The crops in the systems are selected based on their adaptability to the environment, accessibility to water, availability of better seeds, improved technologies, and better market prices. Crop diversification facilitates effective land use, efficient on-farm management systems, optimum water use, effective and efficient integrated pest management, and judicious use of nutrients. Diversification currently aims to increase productivity by the incorporation of alternate vegetables, field crops, fruit crops, flowers, and other options in the cropping system (Modgal, 1998), but ideally, diversification should also aim at improving diet diversity and nutritional balance for the farm families and the local population dependent upon them. Early maturing or bulking potato varieties can fit in the rice or wheat-based cropping system to give maximum productivity and profitability to the system.

Strategies to Mitigate Climate Change Impact

The adverse climate change effect can be reduced by adopting smart practices like inter-cropping, mulching, micro-irrigation, change in planting time, expansion into newly cultivated areas, strengthening climate information along with developing climate resilient varieties — short duration, heat, drought tolerant, disease resistant, and resource use efficient varieties which can be fitted into the cereal-based cropping system.

CIP has identified more than 4,300 varieties of edible potato along with 180 different species of wild potato. While these wild species are too bitter to eat, their importance lies in natural resistance to pests, diseases, and climatic conditions. Till now almost 7,000 potato accessions are safeguarded in CIP's Genebank where research scientists work toward offering the world ever more productive and resilient varieties of potatoes for cultivation. CIP has supplied to Central Potato Research Institute, Shimla, more than 1,100 potato germplasms that have diverse traits such as heat and drought tolerant, late blight resistance, resistant to different viruses, processing quality, short duration, and rich in iron and zinc. The institute utilized some of these germplasms to develop varieties with different characters (Table 1). It can take 10–12 years to release a variety. An ideal potato variety affects not only the yield and quality but also production cost, environmental issues, post-harvest, and yield of future crops.

Climate Resilient Potato Varieties

(a) *Kufri Surya*: It is medium maturing, heat tolerant, and resistant to hopper burn. It has oblong, white smooth skin tubers with pale yellow flesh. This variety is meant for early (September) planting in northwestern plains and for *rabi* and *kharif* plantings in peninsular India. It can also be cultivated in the main crop season in the northwestern and west-central plains. The variety yields excellent smooth and shiny tubers with high proportion of large (> 85 mm) tubers. The reducing sugar content of tubers is less than 100 mg/100 g fresh weight and the dry matter content is 20–21% at harvest (see Fig. 1).

(b) *Kufri Himalini*: It is a medium-maturing potato variety, which is suitable for cultivation in both Indian hills and plains. It is popular in plains and peninsular regions due to its high yield, good keeping quality, and moderate resistance to late blight. It has medium-sized, oval oblong, white tubers, and has a pale yellow flesh color and excellent cooking quality. This variety also has good keeping quality (see Fig. 2).

Table 1. Potato varieties released by CIP and CPRI collaboration in India.

Variety	Characteristics	Parentage	CIP Parent
K. Chipsona 1	Processing	MEX. 750826 x MS/78-79	MEX.750826 (CIP 720124)
K. Chipsona 2	Processing	F-6 x QB/B 92-4	F-6 (CIP 377427.1)
K. Surya	Heat tolerant	K. Lauvkar x LT-1	LT-1 (CIP377257 .1)
K. Chipsona 3	Processing	MP/91-86 (CIP 377427.1) x QB/B 92-4) x K. Chipsona 2	F.6 9CIP 377427.1)
K. Himalini	Late blight resistant	I-1062 x Tollocan	Tollocan (CIP 720054)
K. Frysona	French Fries	MP/92-30 (CIP-378711.5 x AL-575) x MP/90-94 (CIP 378711.5 x MS/78-79)	Muziranzara (CIP-378711.5) and AL-575
K. Chipsona 4	Processing	Atlantic x MP/92-35 (CIP 378711.5 x CIP 720125)	CIP 378711.5 and CIP 720125
K. Lalit	Table	85-P-670 x CP 3192 (CIP 380013.12)	CIP 380013.12
Kufri Lima	Heat Tolerant	C 90.66X C 93.154	CIP 397035.28
K. Thar-2	Water use efficient, virus resistant	CIP92.119 x CIP88.052	CIP 397006.18
Yusimaap	Early maturing, Early bulking red skin, virus resistant		CIP 304351.109

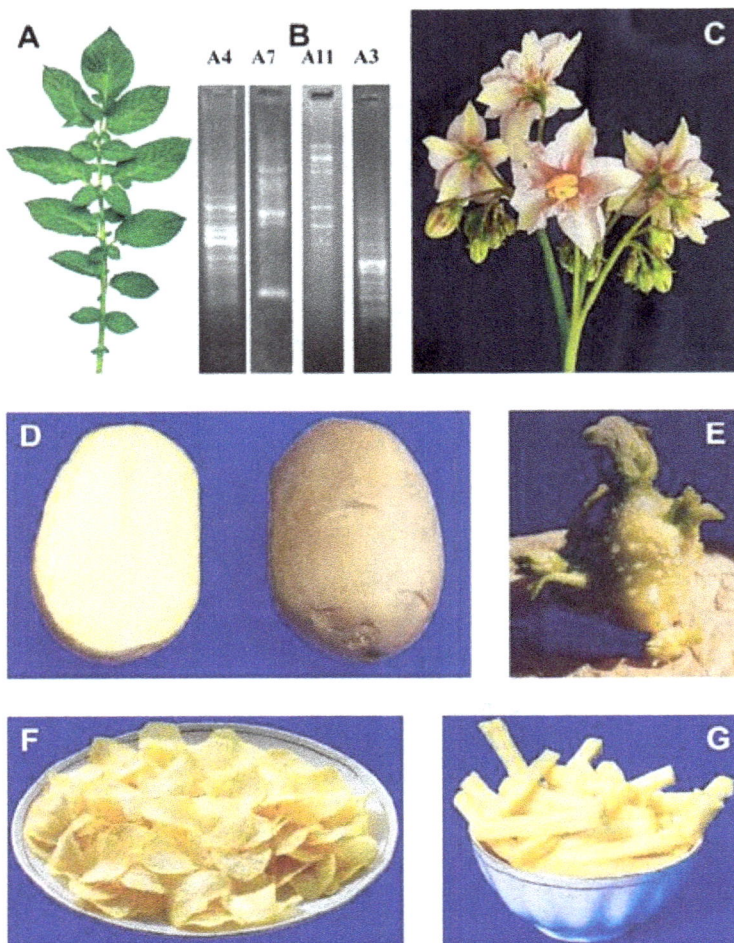

Fig. 1. Morphological and tuber characteristics of K. Surya.

(c) *Kufri Lima*: It is an early maturing, heat-tolerant potato variety with ovoid tubers, white-cream smooth skin, eyes shallow, and pale cream flesh. It has been released from CIP Clone in 2018. It is resistant to PYX, PVY, and is moderately resistant to root-knot nematode. K. Lima is suitable for table purposes as the variety produces attractive tubers without any deformities like cracking or hollow heart. The variety shows luxurious vegetative growth and produces a high tuber yield (38–40 t/ha) with

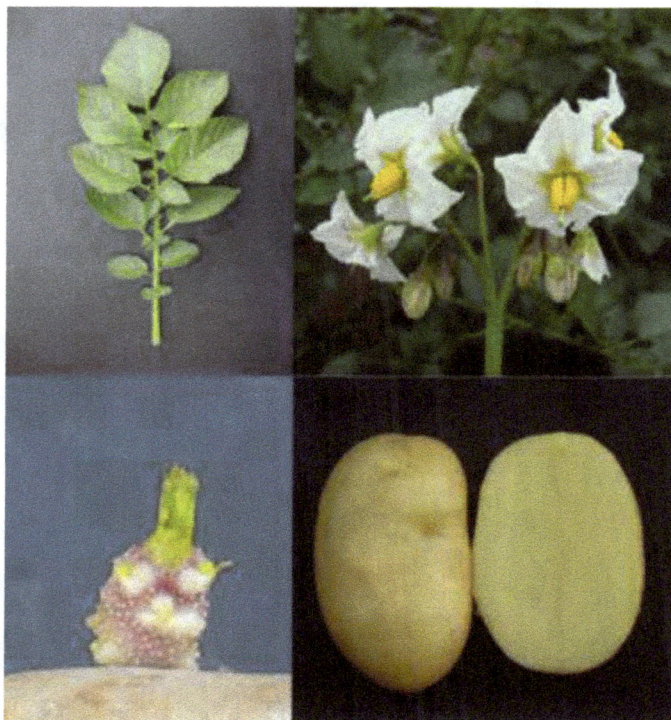

Fig. 4. Morphological and tuber characteristics of K. Thar2.

heat-tolerant clones for early season planting have been imported from CIP-Lima and minitubers were produced last year at Potato Technology Center (PTC), Karnal, Haryana. In 2020–2021, these clones were evaluated in Haryana and CPRI Modipuram to select early heat-tolerant candidate varieties and were also tested for processing quality at CPRI Modipuram, PTC Haryana and PEPSICO (see Fig. 6). These clones were further evaluated in the designated project sites through participatory varietal selection (PVS). The emphasis is on finding suitable new varieties with high dry matter (DM) and low sugar content which will be suitable for processing and could be available in the market in December (see Fig. 7).

(b) *Nutrient-rich Potato*: A total of 57 tetraploid biofortified CIP clones have been imported from CIP Lima to further evaluate its suitability in different agroecological zones of India. This effort

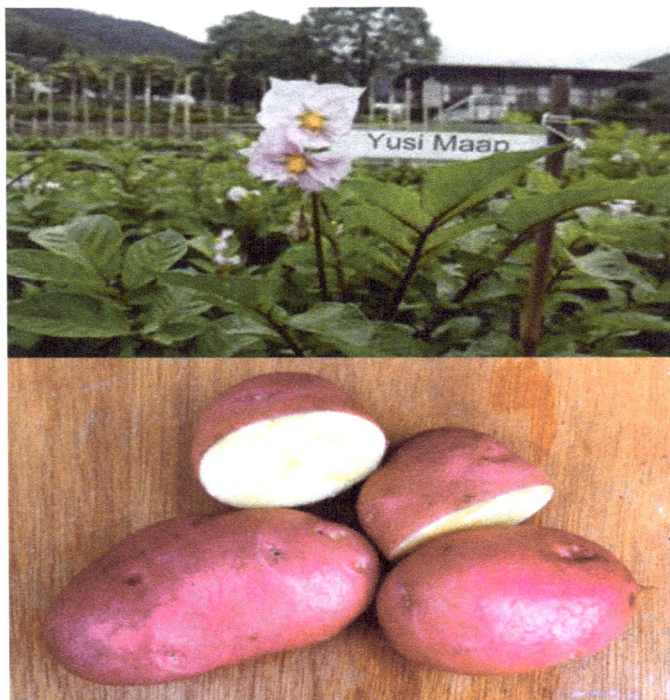

Fig. 5.   Morphological and tuber characteristics of YusiMaap.

will address long-term food security and economic well-being at the country level by providing poor and malnourished people with nutritious potatoes rich in iron and zinc.

Conclusion

The adverse effect of climate change is highly impacting the seed and processing industries. Seed production regions now have become less suitable for quality seed production due to unfavorable weather conditions. Also, unseasonal rain during harvesting is making tubers unsuitable for processing. Smart cultural practices along with the adoption of climate-resilient varieties can reduce the negative effect of climate change on potato production. Besides this, the conservation of germplasms is the key solution to overcome the ill effects of

Fig. 6. Multiplication of Heat Tolerant clones at PTC, Haryana.

Fig. 7. Multiplication of Biofortified clones at PTC, Haryana.

climate change. These germplasms can be utilized for the development of improved varieties. These improved varieties can combat the problems regarding climate change and ensure farmers' income.

References

Bhardwaj, V., Kaushik, S.K., Singh, B.P., Sharma, S., Lal, M., Dalamu, Sood, S., Singh, R., Patil, V., Srivastava, A., Kumar, V., Bairwa, A., Venkatasalam, E.P., Challam, C., Chakrabarti, SK. (2020). Kufri Karan: A multiple disease resistant & high yielding potato variety. *Potato Journal*, 47, 2.

Hijmans, R.J. (2003). The effect of climate change on global potato production. *American Journal of Potato Research*, 80, 271–280.

https://cipotato.org/blog/potato-faces-up-to-climate-change-challenges/.

https://web.inforesources.bfh.science/pdf/focus08_1_e.pdf.

https://www.researchgate.net/publication/273247325.

Jatav, M.K., Dua, V.K., Govindakrishnan, P.M., and Sharma, R.P. (2017). Impact of climate change on potato production in India. In S. Londhe (Ed.), *Sustainable Potato Production and the Impact of Climate Change*. IGI Global, pp. 87–104.

Jennings, S.A., Koehler, A.-K., Nicklin, K.J., Deva, C., Sait, S.M., and Challinor, A.J. (2020). Global potato yields increase under climate change with adaptation and CO_2 fertilisation. Frontiers in Sustainable Food Systems, 4, 519324.

Luthra, S.K., Rawal, S., Gupta, V.K., Bandana, K., Bhardwaj, V., Singh, B.P., Sharma, N., Kadian, M.S., Arya, S., and Bonierbale, M. (2020). Kufri Thar-2: A drought tolerant potato variety. *Potato Journal*, 47, 1.

NHB (2018). Area and production of horticulture crops — All India. https://nhb.gov.in/statistics/StateLevel/2017-18-(Final).pdf.

Pradel, W., Gatto, M., Hareau, G., Pandey, S.K., and Bhardwaj, V. (2019). Adoption of potato varieties and their role for climate change adaptation in India. *Climate Risk Management*, 23, 114–123.

Tiwari, J.K., Luthra, S.K., Dalamu, Bhardwaj, V., Singh, R.K., Buckseth, T., Kumar, R., Gupta, V.K., Kumar, V., Kumar, S., Sood, S., and Kumar, M. Indian Potato Varieties 1949–2020. ICAR-CPRI, Shimla (HP), *eTechnical Bulletin No. 7*, p. 64.

Yadav, S.K., Arya, S., Mohanty, S., Kadian, M., and Luthra, S. (2020). A new heat tolerant potato variety Kufri Lima suitable for early planting in north-western plains of India. *International Journal of Pure & Applied Bioscience*, 8(4), 97–102.

Chapter 7

Small Millets for Food Security in the Context of Climate Change

Satish Kumar Singh*,‖, M.S. Sai Reddy†, R.N. Bahuguna‡,§,
T.V. Anirudh*, and Jai Prakash Prasad*,¶

*Department of Plant Breeding & Genetics, P.G. College of Agriculture,
RPCAU, Pusa (Samastipur), Bihar, India
†Department of Entomology, P.G. College of Agriculture, RPCAU,
Pusa (Samastipur), Bihar, India
‡National Agri-Food Biotechnology Institute (NABI), Mohali,
Chandigarh, India
§Centre for Advanced Studies on Climate Change, RPCAU,
Pusa (Samastipur), Bihar, India
¶Department of Plant Breeding & Genetics, BPS Agricultural College
(BAU, Sabour), Purnea, Bihar, India
‖satish.singh@rpcau.ac.in

Abstract

In the face of escalating climate change impacts on global agriculture, the cultivation and consumption of small millets emerge as a resilient and sustainable solution for ensuring food security. Small millets, comprising crops such as finger millet, foxtail millet, and pearl millet, exhibit exceptional adaptability to diverse climatic conditions, making them well-suited for cultivation in the context of unpredictable weather patterns. It delves into the significance of small millets as a climate-resilient food source. These crops demonstrate remarkable tolerance to heat, water scarcity, and varying soil conditions, thereby mitigating the risks associated with climate-induced disruptions in traditional staple crops. Beyond their adaptability, small millets are nutrient-dense, offering a rich source of proteins, minerals, and vitamins. The promotion of small millets aligns with climate-smart agriculture, contributing to biodiversity conservation and enhancing farmers' adaptive capacity. Furthermore, the cultivation of these millets fosters agroecosystem resilience, as they require fewer inputs like water and synthetic fertilizers compared to major cereals. This emphasizes the pivotal role of small millets in diversifying food sources, enhancing nutritional security, and building climate resilience in agriculture. Embracing these crops can lead to a more sustainable and adaptable food system, offering a practical strategy to navigate the challenges posed by climate change and ensuring food security for vulnerable communities.

Keywords: Millets, Climate change, Sustainable solutions.

Introduction

The group of small millets comprises approximately a dozen crops, including barnyard millet, finger millet, foxtail millet, little millet, kodo millet, and proso millet. These crops are known as "nutri-cereals" and serve as sources of food, feed, and fodder (Saini *et al.*, 2021). They can be grown in a range of environments, including plains, hills, and coastlines, and in various soils and with varying amounts of rainfall. Due to their resilience and ability to withstand drought, small millets are relatively less susceptible to pests and diseases. These crops are unique among cereals for their high content of calcium, dietary fiber, polyphenols, and protein (Hemamalini *et al.*, 2021). Millets contain substantial amounts of amino acids like methionine and cystine and have a higher fat content compared to rice and maize. They also possess nutraceutical properties, such as antioxidants, which have numerous health benefits, including lowering blood pressure, reducing the risk of heart disease and cancer, and decreasing cases of diabetes and tumors. With their potential for contributing to national food security and promoting human health, millets are increasingly gaining attention from food scientists, technologists, and nutritionists.

The estimated global production of millet is 28.4 metric tons per year, with India leading the way at 10.3 metric tons and Africa following closely behind at 8.3 metric tons (Gowri and Shivakumar, 2020). The focus on nutritional security is growing in these regions. In India, millets are grown in many states such as Karnataka, Tamil Nadu, Odisha, Madhya Pradesh, Chhattisgarh, Jharkhand, Andhra Pradesh, Uttarakhand, Maharashtra, Gujrat, and Bihar, covering an area of 1.88 million hectares and producing 2.01 metric tons of which finger millet alone accounts for 60% of the annual planting area and

78% of the production. In Bihar, the area, production, and productivity of small millets are 10,000 hectares, 11,000 tons, and 9.5 quintals per hectare, respectively. Finger millet accounts for over 60% of the area and production of small millets. Despite a reduction in the area, the production has not declined as much due to the increase in productivity levels.

The research on small millets began at Tirhut College of Agriculture, Dholi, Dr. Rajendra Prasad Central Agricultural University, Pusa, Bihar in 1972 with the establishment of the All India Coordinated Research Project on Small Millets. The primary goal of the small millets improvement program is to enhance and stabilize crop production in the millet-growing areas of Bihar. To achieve this objective, it is essential to provide higher-yielding and stable genotypes and to identify improved sources of resistance to various diseases and insect pests in the national germplasm, pre-breeding materials, and other materials from different disciplines. This information can then be used to develop improved varieties.

Types of Small Millets

(a) *Finger millet* (Fig. 1) is processed into flour by milling it with the *testa*, which is high in dietary fiber and micronutrients. The whole meal is used to make traditional foods such as roti (unleavened bread), *ambali* (thin porridge), and *mudde* (dumplings). Regular consumption of whole grain finger millet and its products can reduce the risk of cardiovascular diseases, type II diabetes,

Fig. 1. Finger millet.

Fig. 2. Foxtail millet.

gastrointestinal cancers, and other health problems. The outer layer of the seed coat, which is rich in dietary fiber, minerals, phenolics, and vitamins, is part of the food and provides its nutritional and health benefits.

(b) *Foxtail millet* (Fig. 2) has a higher protein content (12.3%) and iron content (2.8 mg/100 g) compared to rice (6.8% protein and 1.8 mg iron/100 g grain), and it also has a higher fat content (4.3%). The grain is a good source of beta-carotene, the precursor to vitamin A. Foxtail millet is often combined with legumes to make porridge and with soybeans to make mixed flour. With a low glycemic index (GI), it is used in the preparation of low-GI biscuits and a sweet product called burfi, making it a suitable food for people with diabetes. Foxtail millet can also be fermented to produce vinegar, yellow wine, maltose, beer, and other related products. Additionally, it is used to feed cage birds, and the byproduct of foxtail millet can be used as animal feed.

(c) *Proso millet* (Fig. 3) has been used as a staple food. Before rice became the dominant staple in China, it was a widely consumed grain. Among Slavs in Europe, it was one of the main cereals. The countries producing proso millet include China, Russia, India, Eastern Europe, and North America. In Western nations, it has limited economic significance due to the abundance of other cereal crops such as wheat and maize. As a result, proso millet is mostly used as bird feed. However, in recent years, its popularity has grown due to its high-quality proteins.

Fig. 3. Proso millet.

Fig. 4. Barnyard millet.

Proso millet is rich in minerals and vitamins and has similar or better nutritive value compared to common cereals. The protein content ranges from 11.5 to 13%, with a maximum of about 17%. Drying can increase protein levels but also decrease protein quality. The dietary protein in proso millet plays a significant role in cholesterol metabolism. Additionally, it is a good source of potassium, iron, and manganese.

(d) *Barnyard millet* (Fig. 4) is a versatile crop that is grown for both food and animal feed. It is a nutritious source of protein

(11.6%) that is easily digestible and has high dietary fiber content (13.5 g/100g) with a good balance of soluble and insoluble fibers. Its low and slowly digestible carbohydrate content makes it an ideal food for those with sedentary lifestyles. Barnyard millet is highly effective in controlling blood glucose and lipid levels.

(e) *Kodo millet* is a nutritious grain that serves as a great alternative to rice and wheat. With a higher protein, fiber, and mineral content than other staple cereals like rice, kodo millet boasts 8.3% protein and is an exceptional source of fiber, containing 15% compared to rice's 12.9% and wheat's 5.2%. Its flour has a lower gelatinization temperature of 130°C, making it suitable for baking bread and cakes, preparing soups and gravies, making porridge and instant powders, and creating specialty foods using modified flour and starches. Although kodo millet protein is already nutritious, its value can be enhanced by combining it with legume protein. In addition to its nutritional benefits, kodo millet is also high in polyphenols, antioxidants, tannins, phosphorus, and phytic acids. However, these anti-nutrients can negatively affect the solubility and bioavailability of micronutrients like iron, calcium, and zinc. The antioxidant activity of kodo millet decreases when the grain is hulled and cooked.

(f) *Little millet* is high in cholesterol and has the ability to increase the levels of good cholesterol in the body, making it suitable for growing children and providing a boost to overall health (Dey et al., 2022). Its complex carbohydrates are digested slowly, which makes it an ideal food choice for diabetic patients. Little millet is also high in phosphorous (220 mg/100g) and iron (9.3 mg/100g). It is especially beneficial for individuals with low body mass. Some popular dishes that can be made with little millet include *dosa, idli, pongal,* and *kichadi.*

Out of six small millet crops only four crops are grown in Bihar and after intensive research work several varieties of these crops have been developed and released (Table 1). Out of them one finger millet variety RAU 8 has been used as national check for more than five years. Seed rate, sowing time, spacing, and recommended fertilizer dose for different small millet crops have been standardized (Tables 2 and 3).

Table 1. Details of released varieties.

Crops	Varieties	Average yield q/ha	Duration (days)	Suitability
Finger millet	(i) RAU 3	16–18	75–95	Bihar
	(ii) RAU 8	30–35	105–110	India
	(iii) BR407	25–30	115–125	India
	(iv) Rajendra Madua-1	35–40	110–118	Bihar
Barnyard millet	(i) RAU 2	20–25	70–75	Bihar
	(ii) RAU 3	22–30	75–80	Bihar
	(iii) RAU 9	18–20	70–72	Bihar
Proso millet	(i) BR 7	20–22	70–75	Bihar
Foxtail millet	(i) Rajendra Kauni-1	25–30	65–70	Bihar

Table 2. Seed rate, sowing time, and spacing for different small millet crops have been standardized.

Crops	Seed rate (Kg/ha)	Sowing time	Spacing (cm)
Finger millet	10–12	May–August	22.5 × 10
Barnyard millet	08–10	April–July	22.5 × 7.5
Proso millet	08–10	February–April	22.5 × 7.5
Foxtail millet	04–06	April–July	22.5 × 7.5

Table 3. Recommended fertilizers dose for different small millets crops.

Crops	Fertilizer (Kg/ha)		
	Nitrogen	Phosphorous	Potash
Finger millet	40–60	20–30	20–30
Barnyard millet	40	20	20
Proso millet	40	20	20
Foxtail millet	40	20	20

Among breeding trials every year different new cross combinations for resistance against biotic and abiotic stresses are being developed. Breeder and Nucleus seeds are being produced. Apart from varietal development, mini core collections of Finger and Foxtail millets have been developed and published in Field Crops Research in 2011.

Major Diseases of Small Millets

(a) *Blast*: Important disease of pearl, foxtail, barnyard, proso, and little millets and causing heavy yield losses.

Symptoms: Under favorable conditions, spindle-shaped lesions can appear on the foliage and grow larger, merging together and causing the leaf blades, especially from the tip towards the base, to look destroyed. These symptoms can be seen on leaves, the leaf sheath, and stem, and are similar to those found on finger millet, though neck and finger infections are rare. The pathogen responsible for the disease thrives on crop residues and other cereals, and initial infections occur before spreading through air-borne conidia. Disease development is most likely to occur in climates with a minimum temperature of 15–25°C and a relative humidity of more than 85% with intermittent rainfall.

Causal organism: *Pyricularia* spp.

Disease management: Seed treatment with *Trichoderma harzianum* and two sprays of *Pseudomonas fluorescens* at 0.3% at the time of flowering followed by 10 days interval. Use of blast resistant varieties. Spray Carbendazim (0.1%) or Tricyclazole (0.05%) at the first appearance of disease is most effective in minimizing the disease. If the infection persists then repeat the spray after 15 days interval.

(b) *Leaf blight*: Seedling blight or leaf blight of finger millet is a serious and widespread disease, second only to blast in terms of its impact. The disease was first identified by Butler in 1918 as causing foot rot, seedling blight, or leaf blight of ragi in various regions of India.

Symptoms: The emergence of seed rot and seed germination has been linked to the leaf blight pathogen. The initial signs of the disease appear as brown to dark brown spots on the leaf blade. As the disease progresses, the leaf sheath may also display symptoms, particularly in mature plants, where the fungal growth in the center of the lesion can take on a woolly appearance, especially in high humidity environments. Severe infections result in seed discoloration and shriveling. Leaf sheaths in mature plants are particularly susceptible to symptoms, and the woolly growth of the fungus is visible in the center of the lesion, especially in high humidity conditions. The ideal temperature for infection is between 30°C and 32°C. A combination of high humidity and intermittent rains during the emergence of the ear and before grain formation can lead to heavy ear infections and a significant reduction in yield.

Causal organism: Drechslera nodulosum Berk and Curt.

Disease management: Planting crops and turning under the debris can help reduce early infections. Implementing crop rotation with non-host species will also reduce the amount of inoculum. To prevent the spread of the disease, it is recommended to treat seeds with 2.5 gm of Thiram per kg of seed before sowing. At the first sign of disease symptoms on leaves, Dithane M-45 should be sprayed at a rate of 2 g per liter of water. The number of fungicide sprays that may be necessary will depend on the severity of the disease and may range from two to four.

(c) *Cercospora leaf spot*: These leaf diseases have been observed on small millets at all stages of growth, from seedling to grain formation. The first symptoms, which usually start on older leaves, appear as red-brown dots surrounded by a yellow halo. Over time, multiple dots may merge and form larger lesions with a yellow border. In some cases, these lesions can grow to resemble an eye, similar to those caused by blast. This can give the leaves a burnt appearance.

Causal organism: *Cercospora eleusinis*

Disease management: Field sanitation and spraying of Carbendazim @0.1% at 15 days intervals has been reported to reduce infection to some extent.

(d) *Banded leaf and sheath blight*: This disease has been reported in finger, foxtail, barnyard, proso, kodo, and little millets and is becoming a growing problem for all types of small millets. It primarily affects areas with high humidity and frequent rainfall. The disease is characterized by light grey to dark brown oval or irregular lesions on the lower leaves and leaf sheaths. Under favorable conditions, these lesions can quickly enlarge and spread, covering large portions of the sheath and leaf blade. The symptoms produced on various parts of the plant create a distinctive banded appearance, hence the name "banded blight."

Causal organism: *Rhizoctonia solani*

Disease management: To prevent the disease, it is important to ensure proper drainage in the field and to use the recommended amounts of different nutrients. Seeds can be treated with 2.5 gm of Captan per kg of seed before planting. An alternative method of treatment is to use a peat-based formulation of *Pseudomonas fluorescence* at a rate of 16 g per kg of seed, or as a soil application at 7 g per liter of water. Once the crop is established, the lower parts of the plants can be sprayed with Sheethmar (Validamycine) at a rate of 2.7 ml per liter of water.

(e) *Rust*: This disease has been reported to occur in sorghum, pearl millet, and small millets. It can affect the crop at any stage of growth, but the damage is more severe when infection occurs prior to flowering.

Symptoms: Pearl millet rust manifests as round to elliptical reddish-orange pustules on the leaves. Initially, the pustules appear on the distal half of the leaf and then spread to both surfaces. As the pustules mature, they burst and release rusty fragments. When the infection is severe, the plants take on a reddish-brown appearance. Small millet rust is a common problem for foxtail and finger millet, and has also been observed on kodo and little millets.

Causal organism: *Uromyces setariae*

Pearl millet: *Puccinia substriata* var. *indica* (Syn., *P. substriata* var. *penicillariae*)

Kodo millet: *P. substriata;* **Finger millet:** *Uromyces eragrostidis;* **Foxtail millet:** *U. setariae-italiae;* **Little millet:** *U. linearis*

Disease management: To control this disease, it is recommended to grow rust-resistant varieties. Additionally, it is important to remove any collateral hosts and to spray with Propiconazole

(0.1%) or Mancozeb at a rate of 2.5 g per liter of water immediately after the symptoms are observed.

(f) *Smut*: This is the most devastating disease and causes severe losses if infection occurs. It reported on finger millet, foxtail millet barnyard millet.

Symptoms: During the grain formation stage, symptoms of the disease become apparent. The affected ovary transforms into a smut sorus, but it doesn't grow larger than a normal grain. The sori can develop anywhere in the grain, such as on the main rachis, peduncle, or even in finger barnyard millet and foxtail millet. On foxtail millet, the fungus typically affects most of the grains on the ear. The disease is spread through both seed and air. In head smut, the entire panicle becomes a long sorus. The infected panicle may be covered by the flag leaf and not fully emerge. When the membrane covering it bursts, it reveals a black mass of fragments. This disease is primarily spread through seed.

Causal organism:

Grain smut: Finger millet: *Melanopsichium eleusinis;* **Foxtail millet**: *Ustilago crameri;* **Barnyard millet**: *Ustilago panici-frumentacei*

Head smut: Barnyard, Kodo, and Proso millets: *Sorosporium paspalithunbergii*

Disease management: To control the disease, it's recommended to use resistant cultivars, implement proper cultural practices, and apply chemicals as necessary. It was found that early maturing cultivars are more susceptible to grain smut compared to late maturing cultivars. An economical and effective solution is to treat the seed with Carboxin or Carbendazim at a rate of 2 grams per kilogram of seed.

Major Insect Pests of Small Millet Crops

Millets are susceptible to damage from approximately 150 different insect pests throughout their growth and development. Although the current status of each of these pests is not well known, the most economically significant and damaging pests include shoot flies, stem borers, leaf-sucking insects, and insects that attack the panicles. In some regions of India, white grubs are also considered an important pest. In certain seasons, severe yield losses can occur due to sporadic attacks from blister beetles, armyworms, grasshoppers, chinch bugs, leaf beetles, head caterpillars, and head bugs. However, minor pests can also become problematic in some areas due to changes in ecology and agricultural practices.

The major pest across the millets at different stages needs to be tackled by various means as follows:

Seedling pests

(a) *Shoot flies (Atherigona sp., Muscidae: Diptera)*: The shoot fly, belonging to the genus Atherigona, is a pest that affects a variety of millets, as well as crops such as sorghum, maize, and bajra. This pest is relatively uncommon in finger millet. It primarily affects seedlings that are 1–30 days old, causing damage by drying out the central shoot, known as "deadheart." The population of this fly can vary greatly, with low numbers observed from April to June, an increase in July, and peak activity in August. From September, the population gradually decreases and remains at a moderate level until March. Its activity can be impacted by extreme temperatures and prolonged periods of rain.

(b) *Stem borers*: These are destructive group of insects attacking millets. Among them, *Chilo partellus* Swinhoe and *Sesamia inferens* are predominant in India. *S. inferens* infest finger millet in India.

(c) *Ragi stem borer: (Sesamia inferens; Noctuidae: Lepidoptera)*; The pink larvae of the shoot fly bore into the stem, causing damage to the central shoot and leading to "deadheart." The bore holes are visible near the nodes on the stem. The larvae are responsible for creating both deadhearts and stem tunnels. The female fly lays around 150 creamy white, hemispherical eggs that are arranged in two or three rows between the leaf sheath and the stem of the host plant. The eggs take about seven days to hatch. The fully grown larvae measure about 25 mm and are pale yellow with a pinkish-purple hue and a reddish-brown head. The larval stage lasts for 25 days, but in cold months, it may extend to 75 days. The pupation occurs within the larval tunnels.

Sucking pests

(a) *Shoot bug: (Peregrinusmaidis; Delphacidae: Homoptera);* Both the adult forms of the shoot fly (brachypterous and macropterous) and the nymphs suck sap from the plant, reducing its vigor and causing yellowing of the leaves. In severe cases, the younger leaves dry out and eventually, the damage extends to older leaves. The adult shoot fly is yellowish-brown to dark brown with translucent wings. The brachypterous female is yellow, while the macropterous female is yellowish-brown, and the male is dark brown. It lays eggs in groups of 1–4, with 95–100 eggs in total, inside the leaf tissue and covered with a white, waxy substance. The egg period lasts seven days. The nymphal stage goes through five instars in 16 days, with the complete life cycle taking 18–31 days. A heavy infestation during the vegetative stage can cause the top leaves to twist and prevent either the formation or emergence of panicles. The shoot fly is also known to be a vector for transmitting the stripe disease of maize. As a sporadic pest, under favorable conditions, it can produce several generations and lead to complete plant death.

(b) *Aphids (Rhopalosiphummaidis, Melanaphissacchari; Aphididae: Homoptera);* the cornaphid is yellow to dark bluish-green and somewhat ovate. Colonies of aphids are seen in central leaf whorl, stems, or in panicles. The young and adults suck the plant juice. This frequency causes yellowish mottling of the leaves and marginal leaf necrosis. The aphid produces an abundance of honeydew on which molds grow. In panicles, honeydew may hinder harvesting. The aphid also transmits maize dwarf mosaic virus. During rabi, the adult is yellow-colored with dark green legs the colonies are typically found deep inside the plant whorl of the middle leaf on the ventral surface of the leaves, stem, and panicle.

Foliage pests

(a) *Leaf caterpillar*: High incidences of hairy caterpillars, *Aloaalbistriga, Aloalactinea, Spilosoma oblique* north, south and western India have been recorded. Due to gregarious habit and voracious feeding, complete defoliation of millet plants or destruction of seedlings may occur in a short time. They pass the hot summer as diapauses pupae in the soil. Moths emerge about a fortnight after first showers. According to the rainfall distribution, there are one to two generations.

(b) *Cutworms and army worms: (Noctuidae: Lepidoptera);* Caterpillars are defoliators of ragi, maize,bajra and sorghum. *Mythmina separate* has been reported feeding on foliage. The noctuids attacks and defoliate finger millet, *Setaria* sp. and various grasses. The larvae feed on the leaves especially in the nursery. It scraps the green matter of the leaf tissue and the leaves bear a skeletal structure. The young cutworm feeds on plant without cutting off the stems or leaves. Later it begins to cut off foliage. They emerge at night to feed on the roots and shoots of ragi plants and hide in the soil during the daytime.

(c) *Grasshoppers: (Orthoptera);* In India, *Hieroglyphusnigroepletus*, *H. banian, Chrotogonus* spp., *Colemaniaspheneroides* are destructive to millet crops. The nymphs and adults feed on the leaf by consuming large amounts of leaves. They make marginal notchings or holes on the leaves. In case of severe infestation, they defoliate entire leaves and the field gives a grazed appearance. Grasshoppers, locusts feed on foliage and destroy crop.

Management

As a quick solution to controlling this pest, chemical methods have been widely used, particularly on high-yielding varieties and hybrids. However, it is crucial to increase efforts in researching varietal resistance, techniques for mass-rearing natural predators and parasites, and manipulating pest populations through cultural practices. This would allow for an integrated approach to managing this pest in millets in the future. The following are some of the management measures that have been proposed.

Cultural methods

• *Sanitation*: One management measure involves collecting and disposing of stubble and chaffy ear heads, and feeding the stalks to livestock prior to the start of monsoon rains. This helps to decrease the survival of stem borers and midge, thus reducing their numbers

- *Tillage*: Another management measure is deep plowing, which should be done about a month before planting. This exposes the immature stages of insects to the elements and predators, making them more vulnerable and decreasing their populations. Deep plowing is particularly effective in reducing the number of grasshoppers and hairy caterpillars.
- *Cropping pattern*: To minimize damage caused by shoot-fly, midge, and head bugs, it is recommended to synchronize and make early sowings of cultivars with similar maturity over large areas decrease damage caused by shoot-fly, midge, and head bugs. It is advisable to rotate millets with crops such as groundnut or sunflower.
- *Intercropping*: Combining millets with pigeon pea, cowpea, or lablab can also decrease the harm inflicted by stem borers.
- *Weed control*: Consistent and timely weed management in the field can reduce harm caused by *Mythimna separata* and *Spodoptera spp.* An orderly and properly maintained crop is often less inviting to insects, as weeds can serve as a shelter and egg-laying spot for some insects. However, in some cases, plowed fields have been observed to incur greater damage compared to no-till systems. Certain weeds like *Digitaria sp.* and *Echinochloa indica* can provide a habitat for *Spodoptera sp. larvae*, leading to a heightened level of crop damage.
- To enhance plant stand, seedling vigor, and decrease damage caused by shoot-fly, as well as stem borer and sucking pests to some extent, it is advisable to treat seeds with thiamethoxam 70 WS at a rate of 3 g per 1 kg of seeds.
- To minimize damage from shoot-fly, use higher seed rates at 1.5 times the normal amount and delay thinning to maintain optimal plant stand.

Mechanical method

- *Collection and destruction*: The egg masses of hairy caterpillars and other butterfly pests can be gathered manually and eliminated. Aphid outbreaks can also be mitigated by removing and destroying the infected plants. Blister beetles can be manually picked and disposed of. Head bugs and other pests that feed on the ear-head can be removed by flicking them into a container of water with kerosene added.

- *Light traps/pheromone traps/fishmeal traps*: Establish light traps that operate until midnight to monitor, attract, and eliminate adult stem borers, grain midges, and earhead caterpillars.

  - Install sex pheromone traps at 12 per hectare to attract male moths of the species *Helicoverpa sp.* from the flowering to grain hardening stage.
  - Place fishmeal traps, treated with carbamate/organophosphorus insecticides, at 12 per hectare until the crop is 30 days old.

Host plant resistance

In crops like small millets, which have low economic value, the most feasible long-term solution for insect control is to develop high-yielding, resistant varieties and hybrids. Therefore, identifying sources of resistance to major pests is a crucial strategy for the future. This will provide breeders with the material they need to create resistant varieties and hybrids. Emphasis should be placed on multi-pest resistance, i.e., resistance to shoot fly and spotted stem borer. The discovery of sources of resistance to specific pests has expanded the foundation for developing varieties that can effectively address endemic, hot spot, or region-specific pest problems. Research on host plant resistance in ragi is limited, but three varieties (IE 932, 982, and 1037) have been found to exhibit resistance to pink borer. During a resistant screening program, 12 genotypes, including KM 1, RAU 1, RAU 3, Indaf 7, Indaf 8, HR 154, HR 374, HR 1523, PES 110, PES 400, WR 9, and VL 110, were found to be resistant to pink borer. Germplasm screening has shown varying levels of susceptibility to pink borer attacks. Late varieties had a higher incidence of pink borer and grey weevils compared to early and mid-late varieties (Tonapi *et al.*, 2022).

Chemical control

- *Shoot fly (Atherigona soccata (Rondani))*: To manage shoot fly, apply Carbofuran 3G at a rate of 20 kg/ha through furrow application or spray chlorpyrifos 20 EC at a rate of 2 ml/liter of water coinciding with the shoot fly oviposition period (7–14 days after germination) for late-sown crops.

- *Stem borer (Sesamia inferens)*: To manage stem borers, apply Carbofuran 3G granules at a rate of 8–12 kg ai/ha in the whorls at 35–40 days after emergence.
- *Sucking pests*: If the crop is attacked by shoot aphids, spray it with Dimethoate at a rate of 1.5 ml/liter and then spray again 15 days later.
- *Poison bait*: Mix 10 kg of rice bran or wheat bran with one kilogram of jaggery and let it soak overnight. In the evening, mix 100 ml of quinalphos insecticide with the fermented mixture of rice bran or wheat bran, then sprinkle the mixture throughout the field to manage Spodoptera sp.

References

Dey, S., Saxena, A., Kumar, Y., Maity, T., and Tarafdar, A. (2022). Understanding the antinutritional factors and bioactive compounds of kodo millet (paspalum scrobiculatum) and little millet (panicum sumatrense). *Journal of Food Quality*, 1–19.

Gowri, M.U. and Shivakumar, K.M. (2020). Millet scenario in India. *Economic Affairs*, 65(3), 363–370.

Hemamalini, C., Sam, S., and Patro, T.S.S.K. (2021). Awareness and consumption of small millets. *The Pharma Innovation Journal*, 10, 34–37.

Saini, S., Saxena, S., Samtiya, M., Puniya, M., and Dhewa, T. (2021). Potential of underutilized millets as Nutri-cereal: An overview. *Journal of Food Science and Technology*, 58(12), 1–13.

Tonapi, V.A., Ganapathy, K.N., Hariprasanna, K., Bhat, B.V., Amasiddha, B., Avinash, S., and Deepika, C. (2022). Small millets breeding. *Fundamentals of Field Crop Breeding*, 449–497.

Chapter 8

Climate Change Aspects, Water Budgeting, and Operational Agromet Advisory Services

Nabansu Chattopadhyay

*International Society for Agricultural Meteorology
and Agricultural Meteorology Division,
Indian Meteorological Department, India
nabansu.nc@gmail.com*

Abstract

In the next few decades in India, the utilization and management of water in the agricultural sector will be a great challenge due to the ongoing climatic variability and projected climate change. All South Asian countries including India are considered hot regions because of climate change. Thus, more meaningful strategies for water conservation, irrigation, etc., should be given the highest priority. In this chapter, a detailed enumeration is made of observed and projected changes in temperature, rainfall, potential evapotranspiration, and extreme weather conditions. In order to offset the negative impacts of climatic variability and climate change on the availability of water and build resilience to this climate emergency, a number of the major knowledge-based, multi-disciplinary, and participatory interventions like watershed management, water budgeting, alternate wetting and drying, etc., that have been adopted in India in the recent past has been discussed. It has also been shown how the operational Agromet Advisory Services rendered by the India Meteorological Department is delivering these actional information to the farming community to cope and adapt to the impending impact of water crisis.

Keywords: Climatic variability, Climate change, Management of water, Major interventions, Operational Agromet Advisory Services.

Introduction

India is the largest global water consumer. India is globally the second largest producer of wheat and rice, the world's primary food grains (FAOSTAT, 2018). Over two-thirds of the cultivable land in India depends on the monsoons for irrigation. The surface water consisting of rivers and canals facilitates irrigation to the arable land. Of course, the water availability within the country varies, depending on rainfall, groundwater reserves, and river basins. Although India has largely achieved self-sufficiency in the production of food grains, aided by an increase in yields and cultivated areas, climate change and shrinking resources mainly, land and water, have created a huge challenge of feeding the rising billions of population. India is a hotspot for the fallout of the climate crisis. As temperature, rainfall, and other weather patterns become erratic, dependent communities are exposed to vulnerabilities with increased risks of water management in agriculture and failure of traditional coping mechanisms. Climate change will affect the availability, quality, and quantity of water for basic human needs including agricultural sectors. Moors *et al.* (2011) estimated the impact of climate change on the Ganges basin water resources in northern India. Their findings suggest that the increasing annual mean temperature would affect food supply as the depleting water directly suppresses agricultural production.

Because agricultural production is water-intensive, an exorbitant amount of water harvesting may lead to an alarming shrinkage of groundwater level, often manifesting regressive environmental consequences. Water demand is rising exponentially, and among the countries, India is the 13th worst affected in terms of water scarcity. However, with insufficient irrigation water, agriculture growth also appears to serve as a primary source of water scarcity. Both water quality and quantity are serious constraints for productivity growth in agriculture. A rise in water pollution (mainly due to the use of animal manure, chemical fertilizers, and pesticides) also proliferates pressure on water resources. Water resources also have an indirect and possibly time lag impact on climate change. For instance, shortfalls in rainfall can reduce irrigation water supplies, leading to reduced irrigation areas. On the contrary, there would be potentially increased areas under rain-fed crops in the subsequent season with more rainfall. Extreme weather conditions, such as floods, droughts, heat and cold waves, flash floods, cyclones, and hailstorms, are hazardous for crop production.

Some of the major problems of water management in India are flood irrigation, exploitation of groundwater, cultivation of water-intensive crops (for example, sugarcane and rice), insufficient rainwater harvesting structure, waste of water going to seas, and salinity intrusion in coastal districts. The extent of unsustainable use of water calls for the attention of policymakers, businesses, civil society, researchers, and above all, the farming communities. Different strategies are suggested based on available evidence and existing policy preferences by central and state governments. These problems are expected to be aggravated due to ongoing and future climate changes in India.

Existing Problem in Water Use in India: Some Examples

Rice is a water-guzzler because farmers use on average 15,000 liters to produce 1 kg of paddy. No more than 600 liters are needed if proper water management techniques are followed. Given that 45% of the country's total irrigation water is used solely for rice cultivation, the need to improve farming methods is imperative. Besides being wasteful, excessive use of water results in lower yields and adverse environmental effects such as soil salinity and waterlogging. Major problems of water management in India are given in Fig. 1.

Coastal aquifers as vital fresh groundwater resources are subjected to coastal flooding due to storm surges and sea-level rise (SLR). Increasing the seawater intrusion volume (SWIV) from both seaward boundary and land-surface can be expected in coastal aquifers as a result of coastal flooding. Indiscriminate large-scale pumping of groundwater may lead to progressive saltwater intrusion in the freshwater.

In the Ganges Delta using tube wells for water supply can cause serious arsenic poisoning. Arsenic-contaminated water contains arsenous acid (H_3AsO_3) and arsenic acid (H_3AsO_4) or their derivatives.

Flood Irrigation

Exploitation of Ground Water

Cultivation of Water Intensive
Crop (Sugarcane)

Insufficient Rainwater Harvesting
Structure

Waste of water going to seas

Salinity Intrusion in Coastal
districts

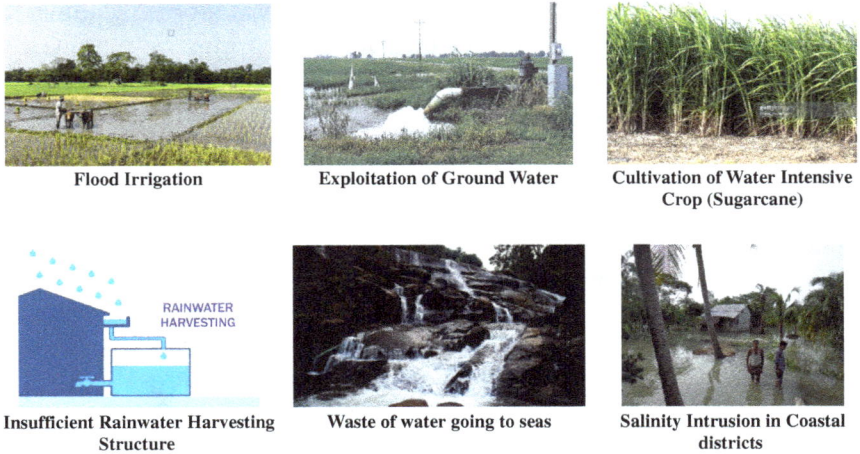

Fig. 1.   Major problems of water management in India.

Floods are also expected to contaminate water sources, destroy water points and sanitation facilities, and therefore pose a challenge to universal access to sustainable water management.

All the existing problems in water use in India are expected to increase by many folds due to the direct and indirect effects of climate change.

Ongoing and Future Climate Change in India

It has been established with a fair degree of accuracy that the climate in India has changed and will change in the future as well. There is a definite shift in temperature in the recent past compared to 1951–1980. The annual mean, maximum, and minimum temperatures averaged over India as a whole show significant warming trends of $0.15°C$, $0.15°C$, and $0.13°C$ per decade, respectively, since 1986. The all-India mean surface air temperature change for the mid-term period 2040–2069 relative to 1976–2005 is projected to be in the range of 1.39–$2.70°C$. Time series of Indian annual mean surface air temperature (°C) anomalies (relative to 1976–2005) from CORDEX South Asia concentration-driven experiments is presented in Fig. 2. On the other hand, spatial patterns of change in surface air temperature (SAT; °C) over India observed changes based on the India Meteorological Department (IMD) dataset is given in Fig. 3. Figure 4

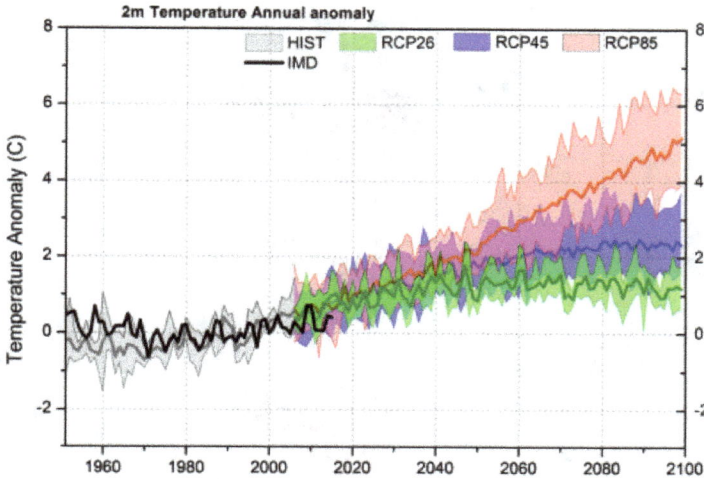

Fig. 2. Time series of Indian annual mean surface air temperature (°C) anomalies (relative to 1976–2005) from CORDEX South Asia concentration-driven experiments. The multi-RCM ensemble mean (solid lines) and the minimum to maximum range of the individual RCMs (shading) based on the historical simulations during 1951–2005 (grey), and the downscaled future projections during 2006–2099 are shown for RCP2.6 (green), RCP4.5 (blue), and RCP8.5 (red) scenarios. The black line shows the observed anomalies during 1951–2015 based on IMD gridded station data.

Source: Krishnan *et al.* (2021).

shows the comparison of maximum temperature changes, minimum temperature changes, and rainfall changes in different agroclimatic zones of India.

Global as well as regional models project an increase in seasonal mean rainfall over India while also projecting a weakening monsoon circulation. Standardized rainfall anomaly of all-India summer monsoon rainfall is given in Fig. 5. However, this weakening of circulation is compensated by increased atmospheric moisture content leading to more precipitation. There is a definite shift in rainfall patterns in the recent past compared to 1951–1980. Frequency of extreme precipitation events may increase all over India, more prominently so over the central and southern parts as a response to enhanced warming. The observed frequency of heavy monsoon rainfall events is presented in Fig. 6. Linear trends in the southwest monsoon rainfall in India based on the data of Indian Meteorological Department is given in Fig. 7. The mean frequency of extreme precipitation for short- and

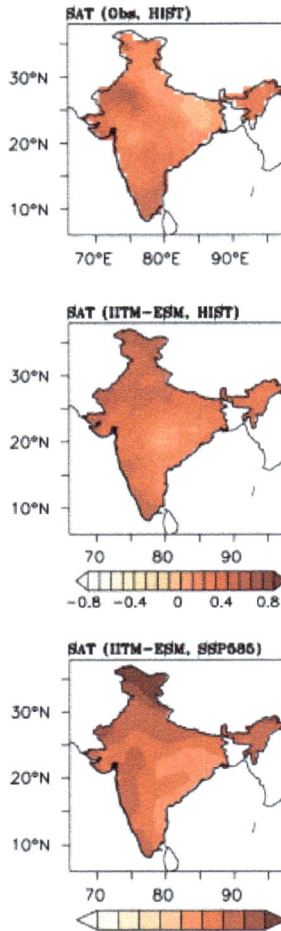

Fig. 3. Spatial patterns of change in surface air temperature (SAT; °C) over India observed changes based on the India Meteorological Department (IMD) dataset. Plots in the middle row are from the IITM-ESM simulations for the historical period, and those in the last row are from the IITM-ESM projections following the SSP5-8.5 scenario. The historical changes in India are shown for the period (1951–2014). Changes under the SSP5-8.5 scenario (last row) are plotted as the difference in mean temperature between the far future (2070–2099) and pre-industrial (1850–1900) periods.

Source: Krishnan *et al.* (2021).

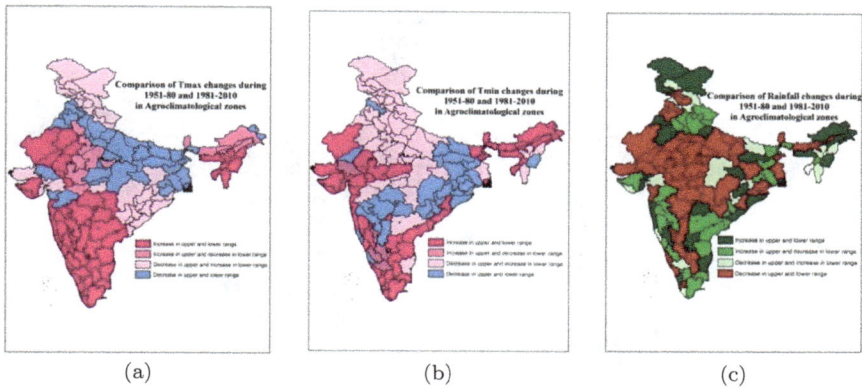

Fig. 4. Comparison of (a) maximum temperature changes, (b) minimum temperature changes, and (c) rainfall changes in agroclimatic zones of India.
Source: Chattopadhyay *et al.* (2019).

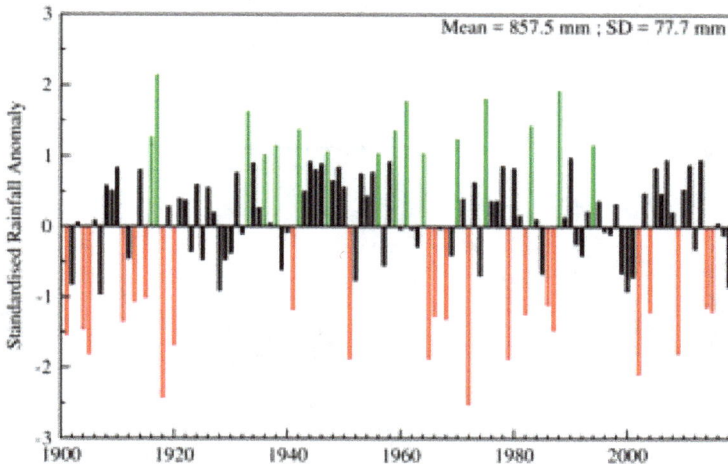

Fig. 5. All-India summer monsoon rainfall.
Source: Krishnan *et al.* (2021).

long-term scale is depicted in Fig. 8. The onset dates are likely to be early or not to change much, and the monsoon retreat dates are likely to be delayed, resulting in lengthening of the monsoon season. The interannual variability of summer monsoon rainfall is projected to increase through the 21st century.

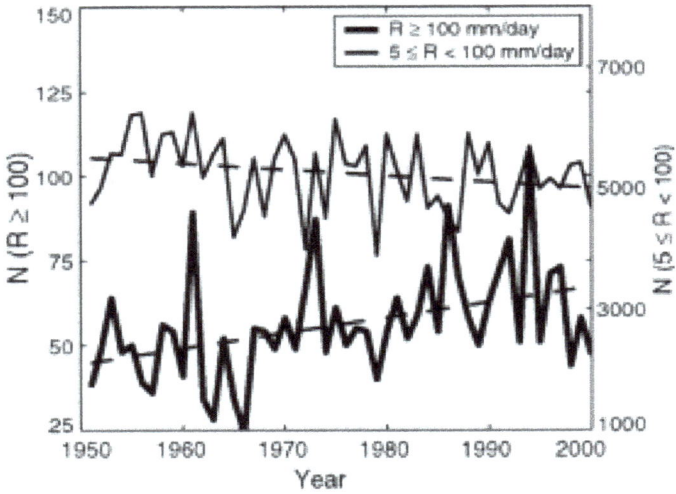

Fig. 6. Observed frequency of a heavy (R 100 mm/day, bold line) and moderate (5 R < 100 mm/day, thin line) daily rain events (Goswami *et al.*, 2006).
Source: Krishnan *et al.* (2021).

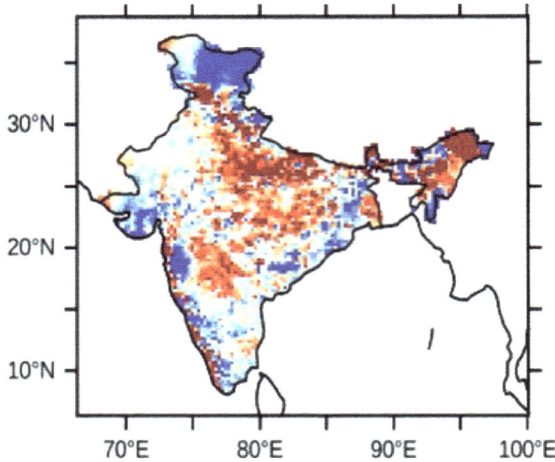

Fig. 7. Linear trends (mm/day over 64 years) in the southwest monsoon rainfall from 1951 to 2015 based on IMD data.
Source: Krishnan *et al.* (2021).

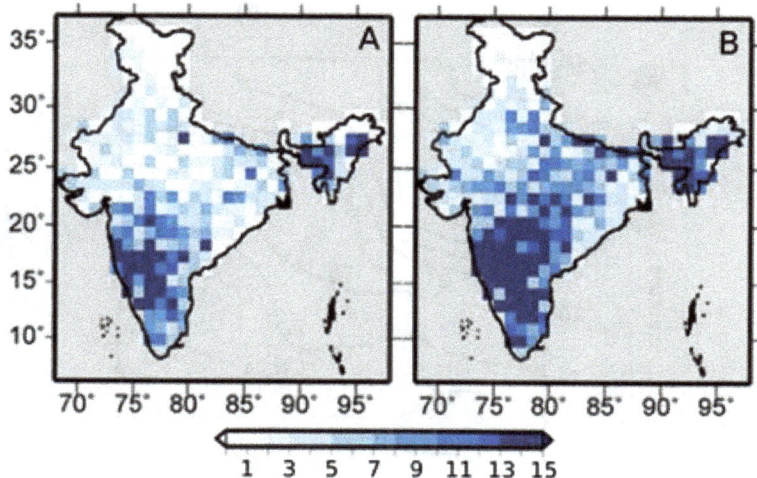

Fig. 8. Multi-model ensemble mean frequency of extreme precipitation for (a) near future and (b) far future.
Source: Krishnan *et al.* (2021).

Future warming seems likely to lead in general to increased potential evapotranspiration over India, although this increase will be unequal between regions and seasons. Such changes could have marked implications for economic and environmental welfare in the country, especially if the increases in evaporation are not compensated by adequate increases in rainfall. One way of assessing interactions between rainfall and potential evapotranspiration is by mapping the number of General Circulation Model (GCM) experiments which yield an increase in the P/PE (precipitation/potential evapotranspiration) ratio. For the monsoon season, all six GCMs agree that the P/PE ratio becomes more favorable over northeastern India, and five out of the six agree that this ratio increases, apart from the extreme south, over the rest of the country. Changes in this ratio are less favorable in the post-monsoon season and in the extreme south of the country. Calculated change in (%) in mean seasonal PE for 1°C of global warming calculated based on different models are presented in Figs. 9–12.

Monsoon (JJAS)

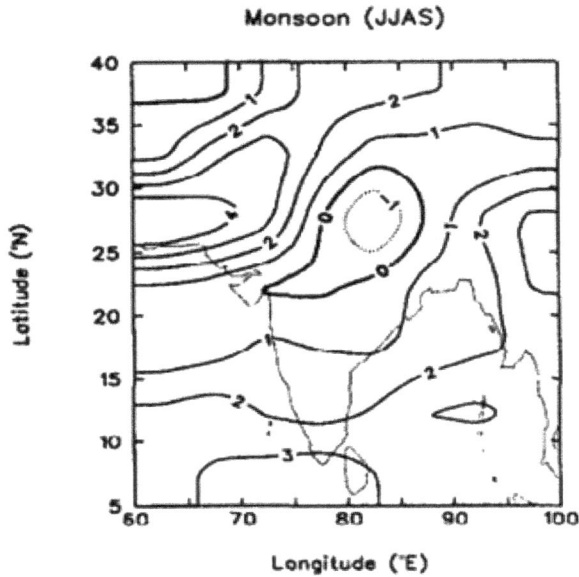

Fig. 9. Calculated change in (%) in mean seasonal PE for 1°C of global warming for the CCC experiment for monsoon (JJAS).
Source: Chattopadhyay and Hulme (1997).

Global warming appears to have increased the intensity of tropical cyclones in the Bay of Bengal. Future projections suggest a likely increase in the number of extremely severe cyclones in response to Indian Ocean warming while changes in frequency remain uncertain. A significant rise (+0.86 per decade) in the frequency of post-monsoon (October–December) season very severe cyclonic storms (VSCS) has been observed in the North Indian Ocean (NIO) during the past two decades (2000–2018). Observations indicate that the frequency of extremely severe cyclonic storms (ESCS) over the Arabian Sea has increased during the post-monsoon seasons of 1998–2018. Climate model simulations project a rise in Tropical Cyclone (TC) intensity and TC precipitation intensity in the NIO basin.

A noticeable increase in flood events has also occurred over the Indian subcontinent (Fig. 13). The climate projections for India also indicate an increase in the frequency of urban and river floods, with an expected rise in heavy rainfall occurrences. Flood propensity is

Monsoon (JJAS)

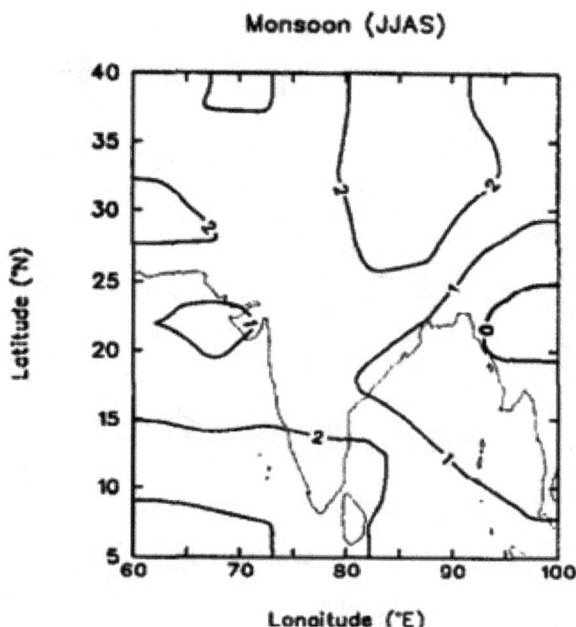

Fig. 10. Calculated change in (%) in mean seasonal PE for 1°C of global warming for the CCC experiment for monsoon (JJAS).
Source: Chattopadhyay and Hulme (1997).

projected to increase over the major Himalayan River basins (e.g., Indus, Ganga, and Brahmaputra) (Fig. 14). The projected enhanced flood risk over India highlights the potential need for better adaptation and mitigation strategies.

Significant drying trends, e.g., Central India, Kerala, and some regions of the south peninsula, also experience higher annual frequency of droughts, with more than two droughts per decade on average for the 1951–2016 period, thus confirming that these regions are becoming more vulnerable to droughts during recent decades. Climate model projections indicate an increase in frequency, spatial extent, and severity of droughts over India during the 21st century (Figs. 15 and 16). One way of assessing interactions between rainfall and potential evapotranspiration is by mapping the number of GCM experiments which yield an increase in the P/PE (rainfall/potential evapotranspiration) ratio. For the monsoon season, all six GCMs

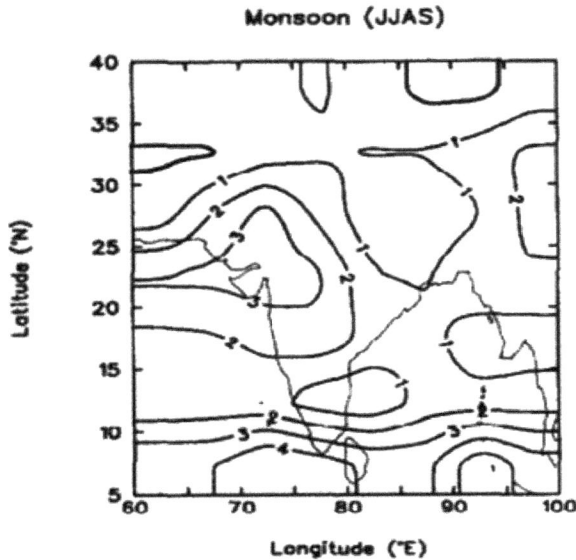

Fig. 11. Calculated change in (%) in mean seasonal PE for 1°C of global warming for the UKTR experiment for monsoon (JJAS).
Source: Chattopadhyay and Hulme (1997).

agree that the P/PE ratio becomes more favorable over northeastern India, and five out of the six agree that this ratio increases,
apart from the extreme south, over the rest of the country (Fig. 17).
Changes in this ratio are less favorable in the post-monsoon season
and in the extreme south of the country. Considering 50-year data, it
is observed that the number of break days during monsoon is higher
in July than in August (Fig. 18).

Sea Surface Temperature (SST) of the tropical Indian Ocean has
risen 1°C on average during 1951–2015 markedly higher than the
global average SST warming of 0.7°C over the same period. During
the 21st century, the SST of the tropical Indian Ocean is projected
to continue to rise. Sea-level rise in the Indian Ocean is non-uniform
and the rate of north Indian Ocean rise was 1.06–1.75 mm per year
from 1874 to 2004 and was 3.3 mm per year in the recent decades
(1993–2015). Steric sea level along the Indian coast is likely to rise
by about 20–30 cm by the end of the 21st century (Fig. 19). Extreme
sea-level events are projected to occur frequently over the Indian

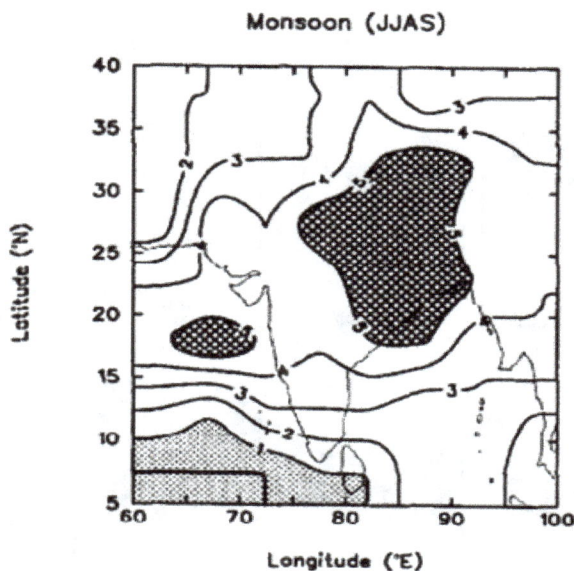

Fig. 12.   Calculated change in (%) in mean seasonal PE for 1°C of global warming for the GCM experiment for monsoon (JJAS).
Source: Chattopadhyay and Hulme (1997).

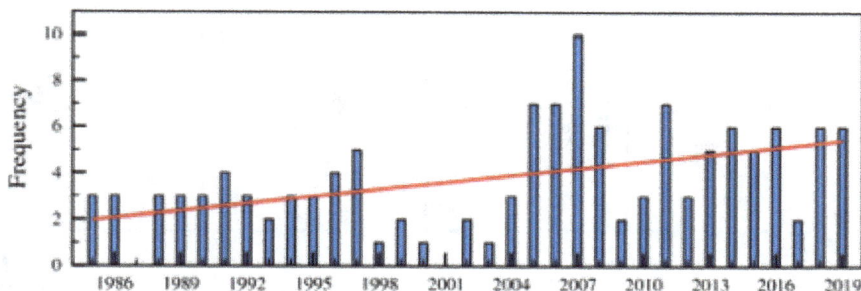

Fig. 13.   Time series of the frequency of severe flood events over India, during 1985–2019 based on flood database of the Dartmouth Flood Observatory (http://www.dartmouth.edu/*floods/Archives/index.html). The red line indicates linear trend.

coast associated with an increase in the mean sea level and climate extremes.

Climate change and increasing demand for water will also put stress on groundwater resources as the availability of surface water

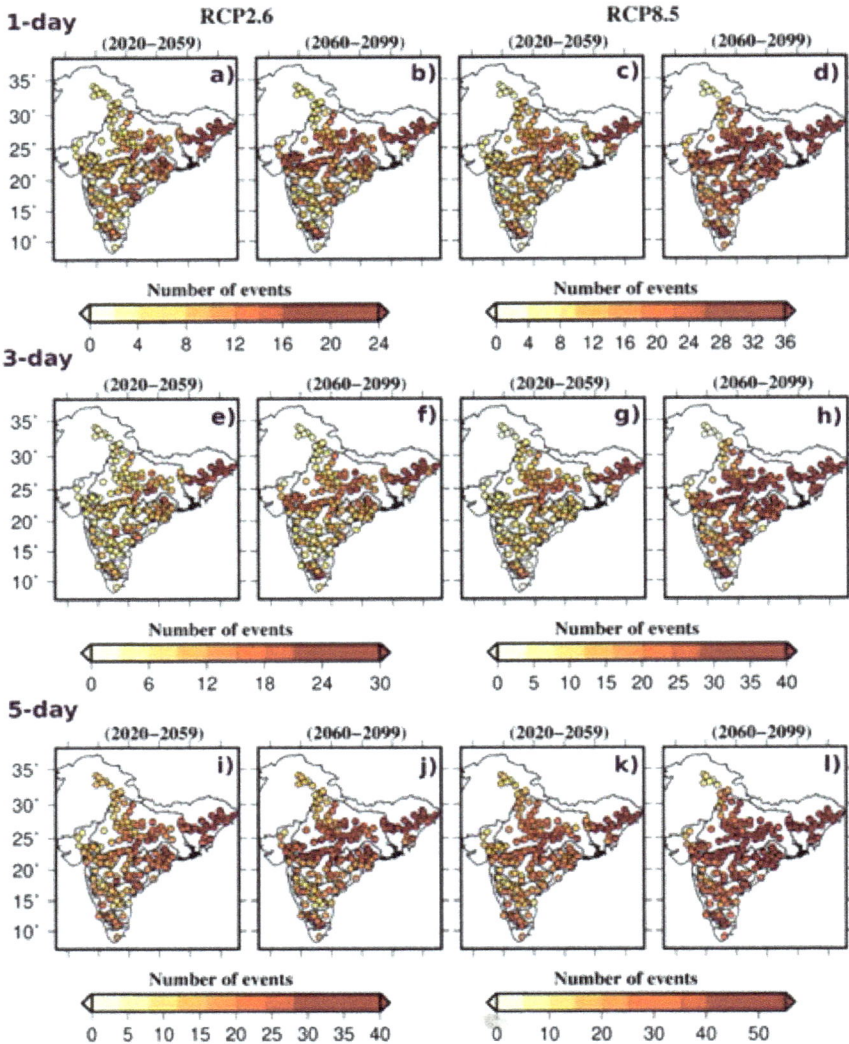

Fig. 14. Changes in frequency of (a–d) 1-day, (e–h) 3-day and (i–l) 5-day duration extreme flood events, projected for (a), (e), (i); (c), (g), (k) near future 2020–2059 and (b), (f), (j); (d), (h), (l) far future 2060–2099, exceeding 20-year return level based on the historic period 1966–2005, as derived from the ensemble mean of five GCMs for (a), (b); (e), (f); (i), (j) RCP2.6 and (c), (d); (g), (h); (k), (l)) RCP8.5 scenario.

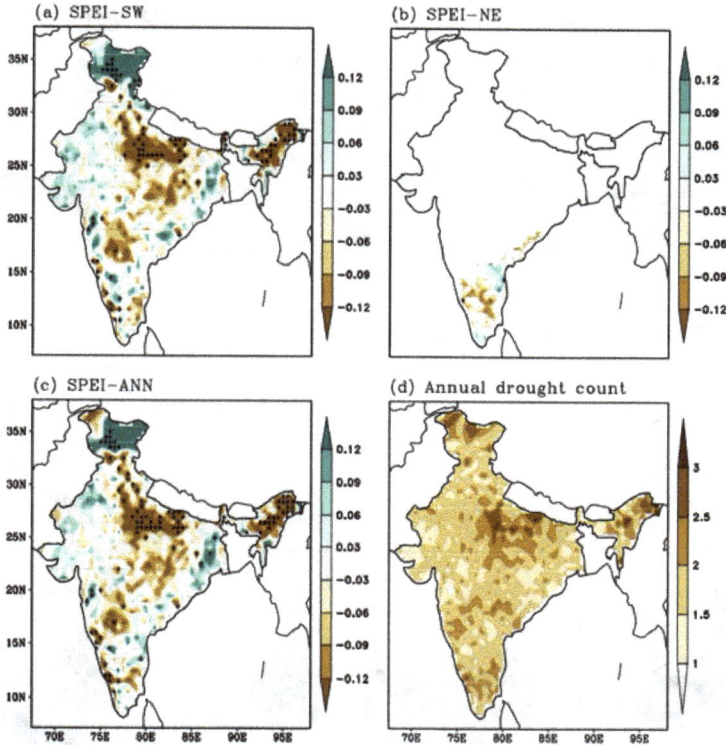

Fig. 15. Spatial pattern of trends (decade−1) in (a) SPEI-SW for JJAS, (b) SPEI-NE for OND, and (c) SPEI-ANN for annual, during 1951–2016. Regions with statistically significant (at 95% confidence level) trends are hatched. (d) Frequency of annual droughts (SPEI-ANN−1.0) per decade, from 1951 to 2016.

is affected by increasing climate variability. Groundwater use could increase by 30% by 2050. The increase in demand for irrigation has already led to severe groundwater stress in some areas, especially in northwestern India. Overuse of groundwater also leads to concentrations of pollutants such as arsenic, iron, manganese, and fluoride, which is a serious concern where groundwater quality is already low, such as in certain locations in India. Floods are also expected to contaminate water sources, destroy water points and sanitation facilities, and therefore pose a challenge to universal access to sustainable water.

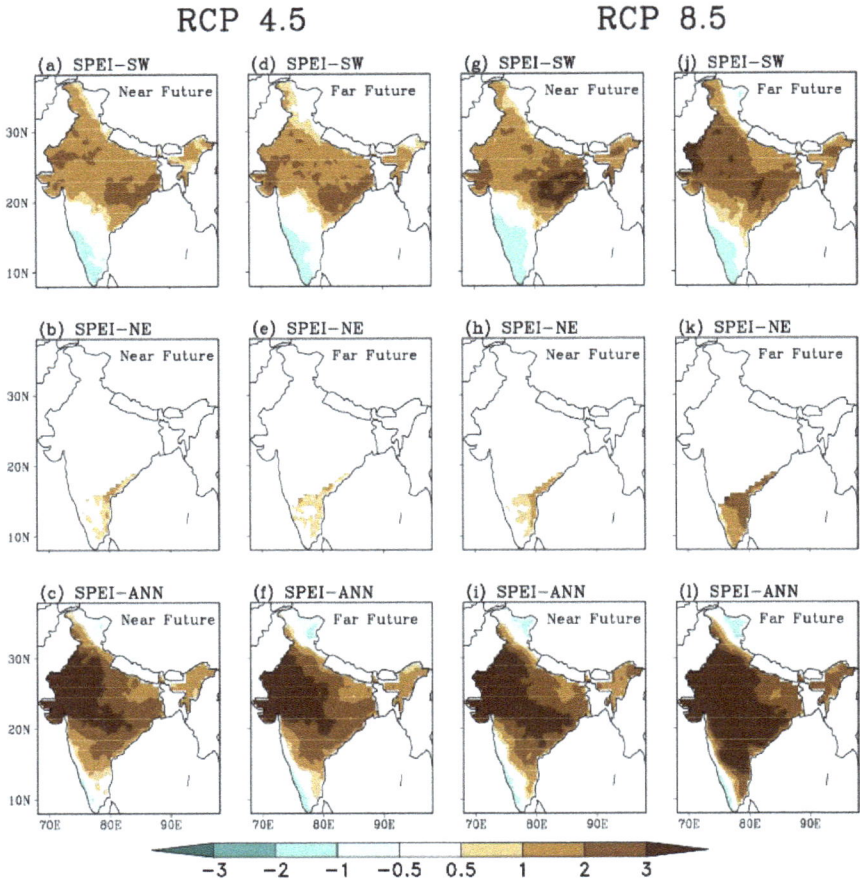

Fig. 16. Changes in frequency (per decade) of moderate and severe droughts (SPEI−1) for (a), (d), (g), (j) JJAS, (b), (e), (h), (k) OND and (c), (f), (i), (l) annual timescale, projected by multi-model ensemble of six CORDEX simulations for (a–c), (g–i) near future (2040–2069) and (d–f), (j–l) far future (2070–2099) from (a–f) RCP4.5 and (g–l) RCP8.5 scenarios, with respect to the historical period (1976–2005).

The importance of sustainable management of water in a changing climate cannot be overemphasized. Depleting water resources, besides land degradation and desertification, loss of biodiversity, and negative impacts of weather variabilities on crop production are

Fig. 17. Number of GCM experiments which indicates an increase in P/PE ratio for each season. The maximum number is six. Areas of agreement in the sign of the change between all six GCMs are shaded.

PERIOD	NUMBER OF BREAK DAYS DURING					
	JULY			AUGUST		
	01–10	11–20	21–31	1–10	11–20	21–31
1888–1917	46	49	53	43	84	26
1918–1947	14	36	21	55	54	25
1948–1977	22	44	64	21	33	41
1978–2003	23	32	39	6	14	37

Fig. 18. Data from the past 50 years show that the number of break days is more in July than August.

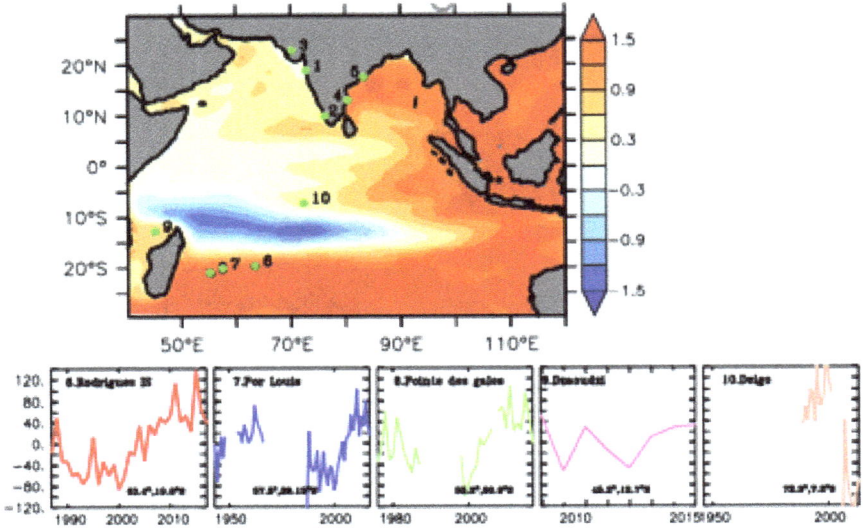

Fig. 19. Spatial map of sea-level trend (mm year−1) in the Indian Ocean from ORAS4 reanalysis for the period 1958–2015 and time series of long-term tide gauge records along the Indian coast and open ocean. The tide gauge locations are marked by green circles. The anomalies are computed with the base period 1976–2005.

direct manifestations of climate change in the agriculture production system. In such a critical scenario, conservation and sustainable management of natural resources, including water, warrants priority action in the policy agenda.

Major Interventions

In order to build resilience to this climate emergency, i.e., climatic variability and climate change, India adopted a knowledge-based, multidisciplinary, and participatory approach involving a wide range of activities toward evidence and policy-focused engagement with all stakeholders. India's approach to the climate crisis especially in terms of the shortage of water seeks to develop knowledge, strategies, approaches, measures, and processes that enable vulnerable communities to cope with and adapt to the impending impacts of the climate crisis in a manner that can be widely adopted, replicated, and upscaled going forward.

To preserve groundwater resources in the coastal zone, it is necessary to manage the threat of seawater ingress due to the rise in sea levels and recurrent floods. Management strategies generally placed the following three categories with the ultimate goal of preserving groundwater resources for current and future use.

(a) Scientific Monitoring, Assessment, and Modeling
(b) Behavioral and Institutional Approach
(c) Engineering Measurement

During flash floods, eroding soil from the hill gets deposited in the agricultural land, resulting in crop loss. Stone bunds and Loose Boulder Structures (LBS) in the gullies were constructed as outlets for the runoff water without affecting the soil. This will conserve rainfall cum runoff and obstruct flash floods and erosion of topsoil from the ridge area to the valley where there are paddy fields. The runoff will be caught by the Continuous Contour Trenches (CCT). The crop fields downstream will have sufficient irrigation through percolation. This is a long-term benefit subject to maintenance of the CCT.

Watershed Development (WSD) in India has been a part of the national approach to improve agricultural production and alleviate poverty in rainfed regions since the 1970s. WSD programs aim to restore degraded watersheds in rainfed regions to increase their capacity to capture and store rainwater, reduce soil erosion, and improve soil nutrient and carbon content so they can produce greater agricultural yields and other benefits.

Water budgeting is a unique approach toward ensuring the optimum, equitable, and most efficient use of water. This involves gaining an understanding of water availability, a community's existing needs and requirements of water, crop planning based on water availability, optimizing irrigation, equitable sharing of water, and considered decisions on groundwater use. Jal Sevaks (water volunteers) from the villages act as representatives to supervise and implement water budgeting activities. Jal Sevaks work as motivators and facilitators, helping village communities implement water stewardship. They are trained to address various challenges in water management. Each Jal Sevak leads the project activities in his own and the neighboring 3–4 villages.

In the past few decades, groundwater use for public supplies, agriculture has increased manifold. Agriculture intensification has resulted in the expansion of groundwater-irrigated areas in India. Village as a whole come together to better understand their water resources so as to enhance the supply through water harvesting structures and manage the use of water.

More crop-per drop' has been the mantra of current public policies around irrigation water. Efficiency savings are always advocated for additional food production for an increasing population. Promotion of micro-irrigation practices through government programs has been localized in a few States — 7.7 million hectares of micro-irrigation, 95% of which is in 10 states.

System of Crop Intensification (SCI) involves soil preparation and management, decreasing crop density per acre and appropriate crop spacing/crop geometry, systematic application of organic inputs and reducing dependence on chemical inputs, spraying of micro-nutrients, and using high-quality seeds. Arrangements are being made to promote the SCI method through plot demonstration during Farmer Field Schools where the farmers are exposed to new farming techniques, field demonstrations, and coping mechanisms within the context of water scarcity and climate variability.

A drainage level of 15 cm is called "safe AWD" because this level will not cause a yield decline. Farmers monitor the water level in the field using a field water tube — a 30-cm length of 15-cm diameter plastic pipe or bamboo, with drilled holes, which is sunk into the rice field until 10 cm of it protrudes above soil level. This has been effective in assuring farmers that the rice plant is accessing water even when there is no standing water in the field.

In traditional rice cultivation, rice is sprouted in a nursery; sprouted seedlings are then transplanted into standing water. With direct seeding, rice seed is sown and sprouted directly into the field, eliminating the laborious process of planting seedlings by hand and greatly reducing the crop's water requirements.

Crop-specific irrigation management practices should be aimed at improving or restoring natural ecosystems. In many high value crops, precision irrigation models and controls like variable-rate drip irrigation and other micro-irrigation systems are gaining wide acceptance including in India. Smart irrigation systems with increased usage of information and communication technology (ICT) and remote sensing. Farmers need to be sensitized about sustainable irrigation water management and the resulting economic and environmental benefits.

A large amount of water is lost in seepage and deep percolation. Loss from deep percolation is estimated at 50% in heavy-textured clay soils and about 85% in light-textured loamy sands and laterite soils. It shows that up to 40% of the water can be saved if farmers adopt proper techniques. Studies in eastern India show effective soil compaction and puddling can reduce percolation losses by 20% and also reduce the risk of crop failure during droughts.

Proper crop-rotation can also reduce water consumption and result in higher yields. In Orissa, the net returns per unit of irrigation water (Rs/cm of water) were estimated at Rs. 28.05/cm of water for rice-mustard-rice rotation, whereas for rice–potato–gingili rotation, it was Rs. 52.81/cm. Better coordination between farmers, scientists, and extension officials is needed to popularize water-saving techniques.

Increased usage of solar pumps has been recommended by policymakers while addressing the challenges arising from "water–energy–food" nexus. He government of India's Kisan Urja Suraksha evam Utthan Mahaabhiyan (KUSUM) is paving the way for installation of standalone off-grid solar pumps for drawing water from surface or underground. Proposed installation of 2 million solar pumps and 1.5 million solarized grid connects to enhance farmers' incomes. However, while promoting solar-based irrigation systems in agriculture sector, we must monitor groundwater extraction to ensure its sustainability.

Livestock — particularly small ruminants — when managed sustainably act as an effective shock absorber for a significant number of vulnerable communities during times of seasonal stress. In semi-arid regions, with low rainfall and agricultural productivity, livestock rearing can be an activity that sustains livelihoods. Livestock rearing — buffalo, goat, sheep, fisheries, poultry and duck, etc. — acts as an "insurance policy" for small and marginal farmers. Livestock also helps in providing a source of protein, supplementing the diets of villagers.

The Government of India uses Earth observation and meteorological data from indigenously built remote sensing satellites for agriculture applications in the country. Fundamental to this governmental ecosystem is the Indian Space Research Organization (ISRO) which builds satellites and payloads and processes the satellite data. Earth Observation techniques are widely recognized in supporting the management of land and water resources and they are nowadays being transferred to operative applications. Satellite-based irrigation advisory system based on dedicated webGIS for farmers and district managers is underway.

Subsidy-based approach to irrigate farmlands has led to negative environmental consequences in India. Over-exploitation of groundwater due to subsidized electricity (it's free) has led to an alarming situation. Diversification with crops like nutri-cereals (sorghum and millets), maize, soybean, fruits, and vegetables, etc., have been suggested to obviate the problem. Adoption, however, will depend on a suitable policy framework with market linkage, creation of supportive infrastructure, and public investments. The World Bank-supported ongoing project titled "Paani Bacho, Paise Kamao" (save water, earn money) could throw practical insights into future public policies to address a very alarming situation. Participatory Irrigation Management (PIM) of resources by farmers' groups may prove to be a better governance model, as demonstrated in some parts of India.

Operational Agromet Advisory Services

The management of weather and climate risks in agriculture has become an important issue due to climatic variability and climate change. The Intergovernmental Panel on Climate Change (IPCC) has highlighted multiple climate risks for agriculture and food security as well as the potential of improved weather and climate early warning systems to assist farmers. Wise use of weather and climate information can help to make better-informed policy and institutional and community decisions that reduce related risks and enhance opportunities, improve the efficient use of limited resources, and increase crop, livestock, and fisheries production. National Meteorological and Hydrological Services (NMHSs) play an important role in providing this weather and climate information to farmers, big and small. However, NMHSs will need realignment, new resources, and training in order to provide location and crop-specific actionable weather and climate services and products that link in available technologies, best practices and go the last mile to reach all farmers (Rathore and

Chattopadhyay, 2016). The Agromet Advisory Services of the India Meteorological Department (IMD) in the Ministry of Earth Sciences is a small step in this direction, aimed at "weatherproofing" farm production.

The sources of weather and climate-related risks in agriculture are numerous and diverse: limited water resources, drought, flooding, early frosts, and many more. Effective weather and climate information and advisory services can inform the decision-making of farmers and improve their management of related agricultural risks. Such services can help develop sustainable and economically viable agricultural systems, improve production, and quality, reduce losses and risks, decrease costs, increase efficiency in the use of water, conserve natural resources, and decrease pollution by agricultural chemicals or other agents that contribute to the degradation of the environment. Thus, the importance of the Agromet Advisory Services that have now been established at district level and block levels in India.

These services meet the real-time needs of farmers and contribute to weather-based crop/livestock management strategies and operations dedicated to enhancing crop production and food security. They can make a tremendous difference in agricultural production by assisting farmers in taking advantage of benevolent weather and in minimizing the adverse impact of malevolent weather.

Today, IMD is implementing operational agrometeorological schemes across the country under a five-tier structure: This structure includes State Agricultural Universities, institutes of the Indian Council of Agricultural Research and Indian Institutes of Technology. Without it, the district Agromet Advisory Services would not be sustainable. An overview of operational Agromet Advisory Services is presented in Fig. 20.

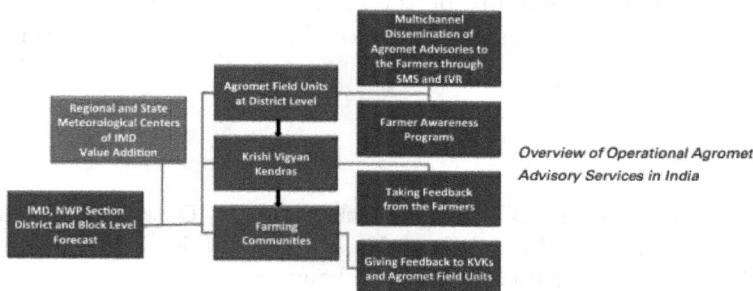

Fig. 20. An overview of operational Agromet Advisory Services.

Such actionable weather information is consistently being delivered to farmers and productivity reports have shown significant increases in yields and with-it food availability and incomes.

As agriculture is weather-dependent at the local level, agrometeorology uniquely combines locale-specific Met-advisories and Agro-advisories that provide timely information including correct irrigation advisories.to farmers so that they can plan their agricultural activities accordingly. It involves advisories that are weather-based, crop and farmer specific, using block-level weather forecasts provided daily by the India Meteorological Department and dynamic crop weather calendars.

Agromet Advisory Services includes all the strategic and tactical decisions from land preparation to harvesting including the choice of crops, etc. All the interventions mentioned aforementioned are communicated to the different users for judicial management of water resources and to save the crops from different climatic risks due to various extreme events.

With increasing weather variabilities, climate change would continue to pose a risk to water availability for agriculture. Focus on sustainable water usage under climate change could be a long-term solution to the challenges of inadequate food and water supplies. The water productivity of major crops in India that have recently been mapped calls for urgent attention to shift areas covered by water-guzzling crops like rice, sugarcane, etc. to other remunerative options. The political economy has to take cognizance of this and repurpose both irrigation and power policies that should incentivize farmers to save water. Research on irrigation practices and technologies, drainage water management, tools for sustainable agroecosystem management, breeding drought-tolerant high-yielding crops, etc., should therefore be the focus of agricultural research systems and incorporated in the operational Agromet Advisory Services.

References

Chattopadhyay, N., Sahai, A.K., Guhathakurta, P., Dutta, S., Srivastava, A.K., Attri, S.D., Balasubramanian, R., Malathi, K., and Chandras, S. (2019). Impact of observed climate change on the classification of agroclimatic zones in India. *Current Science*, 117(3), pp. 480–486.

Chattopadhyay, N. and Hulme, M. (1997). Evaporation and potential evapotranspiration in India under conditions of recent and future climatic change. *Agricultural and Forest Meteorology*, 87, 55–73.

FAOSTAT Statistics database (2018). Food and Agriculture Organization of the United Nations, Rome, Italy. http://www.fao.org/faostat/en/# data.

Goswami, B.N., Venugopal, V., Sengupta, D., Mdhusoodanan, M.S., Xavier, P.K. (2006). Increasing trend of extreme rain events over India in a warming environment. *Science*, 314(5804), 1442–1445. https://doi.org/10.1126/science.1132027.

Krishnan, R., Sanjay, J., Gnanaseelan, C., Mujumdar, M., Kulkarni, A., and Chakraborty, S. (2021). *Assessment of Climate Change over the Indian Region A Report of the Ministry of Earth Sciences (MoES)*, Government of India. https://doi.org/10.1007/978-981-15-4327-2_1.

Moors, E.J., Groot, A., Biemans, H., van Scheltinga, C.T., Siderius, C., Stoffel, M., Huggel, C., Wiltshire, A., Mathison, C., Ridley, J., and Jacob, D. (2011). Adaptation to changing water resources in the Ganges basin, northern India. *Environmental Science and Policy*, 14(7), 758–769.

Rathore, L.S. and Chattopadhyay, N. (2016). Weather and climate services for farmers in India. *WMO Bulletin*, 65(2), 41–43.

https://doi.org/10.1142/9789811296062_0009

Chapter 9

Assessing and Managing Agroclimatic Risks and Opportunities toward Enhancing Food Security: Application of Crop Models

A.K.S. Huda

School of Science, Western Sydney University, Australia
s.huda@westernsydney.edu.au

Abstract

It is imperative to assess and manage agro-climatic risks and opportunities through the application of advanced crop models to enhance food security. As climate change intensifies, agro-climatic factors significantly impact crop yields, challenging the global food production systems. Employing sophisticated crop models provides a systematic and predictive approach to understanding these dynamics. Crop models, utilizing computational algorithms and climate data, offer insights into the complex interplay between environmental variables and crop performance. It underscores the significance of these models in identifying agro-climatic risks, such as extreme weather events and changing precipitation patterns, which directly influence crop productivity. Additionally, these models facilitate the exploration of opportunities by simulating optimal conditions for various crops, aiding in strategic decision-making for crop selection and management practices. The practical application of crop models in assessing agro-climatic risks enables the development of adaptive strategies, including optimal planting dates, irrigation scheduling, and crop rotation plans. By integrating these models into agricultural planning, stakeholders can enhance resilience to climate-induced challenges and optimize resource utilization for sustainable food production.

The pivotal role of crop models is assessing and managing agro-climatic risks and opportunities. Their integration into agricultural practices is essential for fortifying food security in the face of a changing climate, providing a science-based foundation for resilient and sustainable food systems.

Keywords: Climate risk, CSM, Food security.

Introduction

Farmers in developing countries lack access to knowledge and techniques for integrating climate, soil, crop, and agronomic management information to minimize risks and maximize opportunities. The use of an Agroclimatic Risk and Opportunity Management (ACROM) framework (Fig. 1) developed by the author at Western Sydney

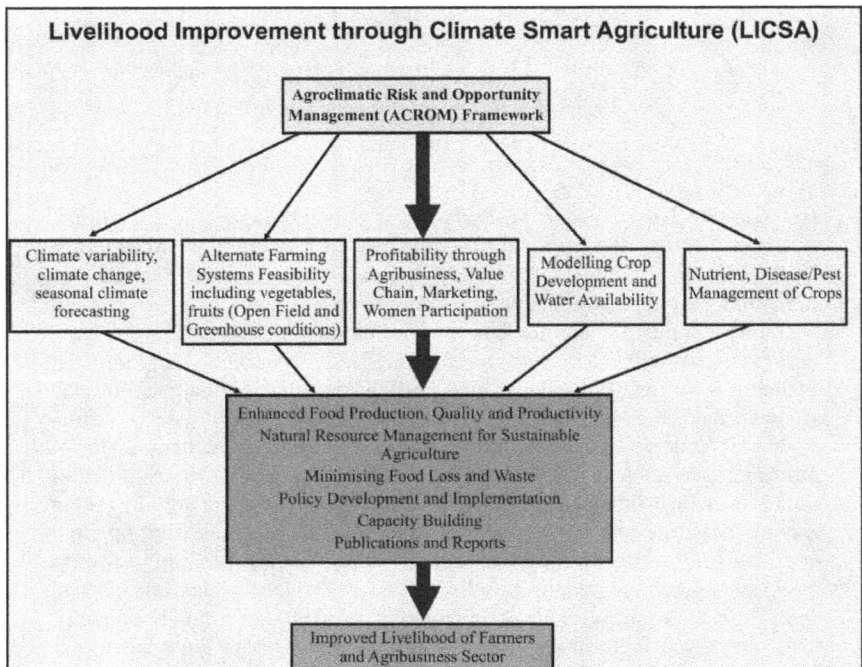

Fig. 1. Agroclimatic Risk and Opportunity Management (ACROM) Framework.

University (WSU) through a multi-institutional research collaboration empowered farmers, researchers, and agribusiness sectors. It helped enhance sustainable farming practices to improve livelihoods and food security through increased quantity and quality of production. ACROM framework was used in a number of research projects supported by highly competitive funding agencies to assist farmers in improving cropping systems feasibility and productivity by:

- Choosing optimal planting times matching crop development period with water availability;
- Selecting crop types and alternate crops;
- Deciding the timing and amounts of fertilizer and irrigation;
- Enhancing resource use efficiency to maximize crop production, crop quality, and minimize crop losses.

This research also helped the farming communities, including the agribusiness sector, to integrate climate, crop, soil, marketing, and management information to enhance food security and promote sustainable natural resource management.

This chapter demonstrates the application of crop simulation models for improved farm decision-making by taking into account the agroclimatic environment. The author shares the insights gained from working in diverse environments through more than 24 national and international research grants. A major benefit of the research has been matching crop development period with water availability: choosing crop types; optimum planting times; timing and amount of fertilizer, water, and pesticide to improve cropping systems feasibility and productivity. Lessons learned from select research projects are shared here.

Improving food security in Qatar: Assessing alternative cropping systems feasibility and productivity in variable climates, soil, and marketing environments (2013–2017)

Although Qatar is a relatively capital-rich country, it is deficient in agricultural resources; food security remains a challenge with the government keen to become less reliant on food imports. The research team provided strategic input into Qatar's Food Security Program.

The research team led by the author emphasized the importance of regular communication with industry and growers throughout the project's lifecycle and delivered extensive industry engagement in the form of media updates, workshops, and regular field visits. It was vital for researchers to spend time in the real-world environment to understand the needs of farmers and other agribusiness stakeholders.

In collaboration with Qatar's Ministry of Municipality and Environment, Department of Agriculture, farmers, and private sector stakeholders, the team used the ACROM framework developed to assess the feasibility of alternate crops such as tomatoes, cucumbers, and squash. Guided by the researchers, farmers and policymakers developed programs to address the lack of information about soil fertility and irrigation water quality for increasing vegetable benefiting farmers and agribusinesses (Huda et al., 2017a; Huda et al., 2017b). One such example of nutrient use efficiency is given in Table 1.

The field-grown squash at the SAIC (Al Sulaiteen Agricultural and Industrial Complex) farm, Doha, Qatar produced a fruit yield of 20t FW/ha ($2.0\,kg/m^2$). Based on published data, the partitioning of nutrients between fruit and above-ground biomass at harvest has been estimated (Table 1).

The nutrient requirements of the above-ground biomass must be met during the growing season but it is apparent that following fruit harvest a large proportion of the nutrients (approximately 60%) remains in the above-ground crop residues. If these residues are conserved in the field (not burnt or removed for animal feed or compost), they provide a valuable source of nitrogen and potassium for the following crops. Assuming no off-site losses of the nutrients in

Table 1. Partitioning of nutrients in the above-ground biomass of field-grown squash with a yield of 20t (FW)/ha ($= 2\,kg/m^2$).

	Nitrogen (kg/ha)	Phosphorus (kg/ha)	Potassium (kg/ha)
Fruit	84	9	100
Leaf+stem at harvest	116	13	155
Total above-ground biomass	200	22	255

Source: https://www.ipni.net/app/calculator/crop/SQ; Huett and Dettmann (1992).

the crop residues, it is estimated that approximately 120 kg N/ha (60% of residue N), 14 kg P/ha (66% of residue P), and 240 kg K/ha (100% of residue K) will probably be available for uptake by the next crop. This nutrient budgeting approach can be used to inform nutrient inputs for the next crop; combined with soil test information on current soil phosphorus and potassium fertility, fertilizer recommendations can be made (Huda *et al.*, 2017b).

The project initiated a new dialogue between government and private sector stakeholders in Qatar on food security strategies. Some findings included the recommendation of the AquaCrop model (Raes *et al.*, 2012) for use in simulating yields of crops to assess the effects of agronomic practices including planting dates, soil characteristics, and water/nutrient use efficiency. The research also published evapotranspiration data analysis for Qatar (Issaka *et al.*, 2016).

Safeguarding food and environment in Qatar (SAFE-Q) (2014–2018)

The food supply chains distribute the food demanded by the consumers; however, wastes occur both in operations and at the end of the chain, in the consumption stage. Some of this waste is unavoidable, but there are opportunities to reduce certain types of waste by improving operations and changing consumption behaviors. Policymakers from the Qatar National Food Security Program and the Department of Agriculture were engaged intensively in developing the food security decision-making framework that met the Qatar Government's requirements for strategic assessment of possible food-supplying locations. The combination of Qatar Government's on-the-ground capacity, the research expertise of the author in Climate-Smart Research coupled with the advice of co-researchers from other institutions ensured the project realized impact in a challenging area of food security (Aktas *et al.*, 2018; Irani *et al.*, 2018).

Food security and climate change: Evaluating mismatch between crop development and water availability in the Asia-Pacific Region (2010–2014)

The research team used the ACROM framework to improve Asia-Pacific farmers' understanding of the impacts of climate variability on key crops by educating them to adapt to new techniques, grow new varieties of crops, and use proper natural resources. A schematic diagram of Livelihood management through Climate smart Agriculture (LICSA) based on Agroclimatic Risks and Opportunity Framework (ACROM) is presented in Fig. 1. The research program enhanced the capacities of the local farmers and extension officers. For example, "Agromet" Advisory Services in Andhra Pradesh (India) used mobile phone-based messaging to disseminate information directly to farmers growing groundnuts. As a result, groundnut productivity remained resilient despite increased weather variability due to climate change. In 2016, an additional 175 farmers were taught how to pick the ideal week to sow their groundnuts.

Farmers praised the framework for helping them achieve better crop yields. Our research drove the Chinese Government to establish a Disaster Fund to deal with extreme weather events as part of an adaptation strategy (Huda *et al.*, 2011, 2012).

The effects of climate change on pests and diseases of major food crops in India, Bangladesh, and Australia (2009–2011)

The Asia-Pacific Network for Global Change Research funded an interdisciplinary research team in member countries (Australia, India, and Bangladesh), to develop the first comprehensive analytical model (Late Blight Prediction System) using ACROM framework.

Potato consumption has grown strongly in developing countries, and potatoes have become a valuable cash crop. In India and Bangladesh, however, Potato Late Blight (PLB) disease causes severe crop losses, often pushing farmers into poverty and sometimes to suicide. Using the ACROM framework, the research team developed the first comprehensive analytical model for assessing as well as predicting the impact of PLB. The Late Blight Prediction System also facilitated analysis of PLB severity by providing growers with the ability to forecast disease occurrence, reduce unnecessary fungicide applications, and schedule control treatments. When the impact of future climate change on PLB was projected and compared to historical records for the 1981–2010 seasons, the model demonstrated future patterns and severity of PLB occurrence The framework and projections allowed farmers to resolve existing issues of PLB and be prepared for the future (Luck *et al.*, 2012).

Climate and crop disease risk management: An international initiative in the Asia Pacific region (2006–2009)

An international network of researchers and a collaborative plan for advancing the objectives of the APN were promoted in countries including Cambodia, Laos, Vietnam, Bangladesh, and India (Huda *et al.*, 2007). The activities and output included:

- Workshop engaging project team with local stakeholders;
- Focus on "what do the users want" rather than "what does the modeller want" approach;
- Engagement with the users to assess the economic, social, and environmental outcomes of disease management interventions (Banerjee *et al.*, 2010; Sarmah *et al.*, 2009);
- Further research efforts building on the APN seed funds outcomes.

Impact assessment of climate variability and climate change on crop water productivity of wheat at selected Indian and Australian locations: A crop growth simulation approach (2016–2017)

The project evaluated the impact of climate variability to reduce the gap between attainable crop yield and actual yield obtained by farmers. During the last 30 years, the average gap varied between 33% and 139% across the West Bengal state. This project assessed the historical and forthcoming (2021–95) changes in climatic parameters (temperatures, rainfall, solar radiation) and their influence on the yield and water-use pattern of wheat in selected locations. These include West Bengal State in India (Jalpaiguri, Malda, Murshidabad, Nadia, Birbhum, and South 24 Parganas districts) and Australia (Dalby in Queensland; Junee, Trangie in New South Wales; Esperance, Jerramungup in Western Australia).

Daily rainfall, maximum temperature (Tmax), and minimum temperature (Tmin) data were analyzed for a period of 1983–2012 (Indian locations) and 1900–2013 (Australian locations). Future climatic dataset under RCP 8.5 scenario for a period of 2021–2095 was used for both countries. Potential yield and attainable yield were simulated for both past and future climatic conditions by using Decision Support System for Agro-technology Transfer (DSSAT v4.5) model. Simulated yields were compared with actual yields obtained by farmers (Mukherjee *et al.*, 2017; Mukherjee and Huda, 2018). Outcomes of the project encouraged district authorities to mobilize resources to minimize the yield gaps by promoting improved agronomic and sustainable natural resource management practices leading to new projects on the livelihood improvement of farmers through climate-smart agriculture.

Livelihood improvement through Climate-smart agriculture: An Australia–India Initiative (2017–2021)

Engagement with farmers by the research team led by the author ensured education and knowledge exchange occurred at the end-user level. The integration of the impact process into the research trajectory from the conceptual stages enabled end-user uptake and practice change at the appropriate time. Work plan was developed through workshops by engaging with local farmers/farming and rural communities, researchers, government agencies, policymakers, and NGOs in the Birbhum district of West Bengal. The research efforts enhanced livelihood by capturing excess rainwater during the rainy season through climate-smart agriculture. The stored water was reused in particular to introduce new high-value cash crops/enterprises in post-rainy season resulting increased agricultural crop yields and income. It provided capacity building for local farmers through on-farm testing and demonstration of new climate-smart technologies (Huda *et al.*, 2019).

The team successfully negotiated significant co-funding from Birbhum District Authority, Government of West Bengal, to support the project activities. It arranged two major workshops in Australia (March 2018 and August 2019) and two in India (December 2017 and February 2019)) and were attended by key people from India and Australia including the High Commissioner of India to Australia and senior Australian DFAT officials responsible for India.

Matters arising from these workshops were directly fed into a comprehensive report — An India Economy Strategy 2035 — commissioned and launched by the Australian Prime Minister. The report of the meeting was released by the Australian Deputy Prime Minister at the Parliament House. Key outcomes included are as follows:

- Improvement in the livelihood of Indian farmers and their families;
- Increased water resource status of the selected village/site;
- Enhanced crop water use efficiency;
- Increased cropping intensity during pre and post-rainy seasons with assured harvested water;
- Nutrition enrichment for the villagers with fruit cultivation;
- Employment opportunities for the villagers during the lean period;
- Improved socioeconomic status as a Model for other farmers to emulate;
- Improved bilateral cooperation.

Lessons Learned for Future Research

- Undertaking assessment of the robustness of the available agromet-related products and services and identifying its limitations;
- Institutional mechanism to reach forecasts and advisories to end-users for decision-making in agriculture;
- Coordinating with relevant government agencies involved (including relevant Bureau of Meteorology and Department of Agriculture/Primary Industries) in the generation and application of agromet;
- Motivating cooperators in developing further ideas and tools for sustainable crop rotations;
- Producing guidelines for developing and applying the tools of hard systems. Farming Systems Research (FSR) showed that most guidelines are produced with a narrow focus.

The researchers that develop hard systems tools have largely ignored the guidelines with broader applications and possibly greater impact, such as those concerned with farmer participation in system design and application (Robinson *et al.*, 2007). The following are the key ingredients for effectiveness and impact:

- Respect the farmer's perspective of the situation and decision-making process;
- Understand the decision-making process;
- Appreciate the researchers' motivations and biases.

There is obviously a lot yet to be learned. Integrating our understanding of the technical world of tools with the mental world of decision-making is a challenging task. Our extensive research and research leadership should include the initiation and supervision of client-focused, industry-relevant research and development. Research emphasis should be given to the development and delivery of agricultural systems analysis, improved risk management strategies, applied climate forecasting, and climate variability assessments for food security. Much of this effort should involve working directly with stakeholders including other key scientists, policymakers, government agencies, farmers, and agribusiness. Expertise gained from key researchers around the world combined with strong interest and deep commitment should ensure that research outcomes are directly useful to end users.

Acknowledgments

The author is supported by highly competitive external funding bodies including Qatar National Research Fund (QNRF), Australian Centre for International Agricultural Research (ACIAR), Australia's Grains Research and Development Corporation (GRDC), Asia-Pacific Network for Global Change Research (APN), Australia–India Council, Australia Endeavour Fellowship Program, The Crawford Fund, Australian Agency for International Development, and Australia's Land and Water Resources Research and Development Corporation.

References

Aktas, E., Sahin, H., Topaloglu, Z., Oledinma, A., Huda, A.K.S., Irani, Z., Sharif, A.M., van't Wout, T., and Kamrava, M. (2018). A consumer behavioural approach to food waste. *Journal of Enterprise Information Management*, 31(5), 658–673.

Banerjee, S., Bhattacharya, I., and Huda, A.K.S. (2010). Weather sensitivity and downy mildew and alternaria blight of mustard in the Gangetic West Bengal, India. *Journal of Science Foundation*, 8(1 & 2), 77–81.

Huda, A.K.S., Hind-Lanoiselet, T., Derryl, C., Murray, G., and Spooner-Hart, R.N. (2007). Examples of coping strategies with agrometeorological risks and uncertainties for Integrated Pest Management, Chapter

16. In Sivakumar and Motha (Eds.), *Managing Weather and Climate Risks in Agriculture*. Springer, pp. 265–280.

Huda, A.K.S., Issaka, A.I., Kaitibie, S., Haq, M.N., Abdella, K., Moody, P.W., Moustafa, A.T., Goktepe, I., Coughlan, K.J., Pollanen, M., and Vock. N. (2017a). Assessing potential yields of selected vegetables and evaluating alternate management practices to improve Qatar's food security. *Chronicle of Bioresource Management*, 1(1), 016–019.

Huda, A.K.S., Issaka, A.I., Kaitibie, S., Haq, M.M. , Goktepe, I., Moustafa, A., Abdella, K., Pollanen, M., Moody, P.W., Vock, N., Huda, N., and Coughlan, K.J. (2017b). *Improving Food Security in Qatar: Assessing Alternative Cropping Systems Feasibility and Productivity in Variable Climates, Soil and Marketing Environments (NPRP 6-064-4-001). Final Report*. Qatar National Research Foundation, p. 79.

Huda, A.K.S., Sadras, V., Wani, S., and Mei, X. (2011). Food security and climate change in the Asia-Pacific region: Evaluating mismatch between crop development and water availability. *International Journal of Bioresource and Stress Management (IJBSM)*, 2(2), 137–144.

Huda, A.K.S, Sarkar, N.C., Mukherjee, A., Singh, B., and Mohangandhi, P. (2019). Australia–India collaboration for improving farmers' livelihood through climate-smart agriculture. *Proceedings, National Seminar on "Sustainable Resource Management for Enhancing Farm Income, Nutritional Security and Livelihood Improvement"*. Visva-Bharati University.

Huda, A., Spooner-Hart, R., Murray, G., Hind-Lanoiselet, T., Ramakrishna, Y., Desai, S., Thakur, R., Chattopadhyay, C., Jagannathan, R., and Khan, S. (2005). Climate related agricultural decision making with particular reference to plant protection. *Journal of Mycology and Plant Pathology*, 35, 513.

Huda, A.K.S., Wani, S.P., Mei, X., and Sadras, V. (2012). Food security and climate change in the Asia-Pacific region: Evaluating mismatch between crop development and water availability. *APN Science Bulletin*, 2, 42–48.

Huett, D.O. and Dettmann, E.B. (1992). Nutrient uptake and partitioning by zucchini squash, head lettuce and potato in response to nitrogen. *Australian Journal of Agricultural Research*, 43(7), 1653–1665.

Irani, Z., Sharif, A., Lee, H., Aktas, A., Topaloglu, Z., Wout, T., and Huda, S. (2018). Managing food security through food waste and loss: Small data to big data. *Computers and Operations*, 98, 367–383.

Issaka, A., Paek, J., Abdella, K., Pollanen, M., Huda, A.K.S., Kaitibie, S., Goktepe, I., Haq, M.M., and Moustafa, A. (2016). Analysis and calibration of empirical relationships for estimating evapotranspiration in Qatar: A case study. American Society of Civil Engineers (ASCE). *Journal of Irrigation and Drainage Engineering*, 143(2), 1–7.

Luck, J., Asaduzzaman, M., Banerjee, S., Bhattacharya, I., Coughlan, K., Chakraborty, A., Debnath, G.C., De Boer, R.F., Dutta, S., Griffiths, W., Hossain, D., Huda, S., Jagannathan, R., Khan, S., O'Leary, G., Miah, G., Saha, S., and Spooner-Hart, R. (2012). The effects of climate change on potato production and potato late blight in the Asia-Pacific region. *APN Science Bulletin*, 2, 28–33.

Mukherjee, A. and Huda, A.K.S. (2018). Assessment of climate variability and trend on wheat productivity in West Bengal, India: Crop growth simulation approach. *Climatic Change*, 147(1–2), 235–252.

Mukherjee, A.K., Huda, A.K.S., Thentu, T.L., and Banerjee, S. (2017). Increasing wheat productivity under variable and changing climatic conditions in West Bengal, India. *International Journal Bioresource Management*, 8(3), 473–476.

Raes, D., Steduto, P., Hsiao, T.C., and Fereres, E. (2012). *AquaCrop — the FAO Crop Model to Simulate Yield Response to Water. Reference Manual Version 4*. FAO Land and Water Division, Rome, Italy.

Robinson, J.B., Freebairn, D.M., Huda, A.K.S., and Rattray, D. (2007). Improving agricultural sustainability and profitability via the use of computerised decision support systems is challenging and complex. In A. Ahmed (Ed.), *Knowledge Management and Sustainable Development in the 21st Century: World Sustainable Development Outlook 2007*. Greenleaf Publishing Limited, Sheffield, pp. 30–140.

Sarmah, K., Banerjee, S., and Huda, A.K.S. (2009). Simulating the impact of climate change on the performance of rapeseed and mustard through crop simulation model. *International Journal of Agriculture, Environment and Biotechnology*, 2(4), 318–322.

Chapter 10

Extreme Weather Events: Basics, Agricultural Impacts, and Protective Measures

Y.S. Ramakrishna*, G.G.S.N. Rao, and A.V.M. Subba Rao

AICRPAM, CRIDA, Hyderabad, India

*ramakrishnays@gmail.com

Abstract

This chapter delves into the fundamentals of extreme weather events, their profound impact on agriculture, and the imperative for implementing protective measures to safeguard food production. Extreme weather events, including floods, droughts, hurricanes, and heat waves, are increasingly frequent and intense due to climate change, posing substantial threats to global food security. The agricultural sector is particularly vulnerable to these events, with disruptions ranging from crop failures and livestock losses to soil degradation and altered growing seasons. This abstract emphasizes the need for a comprehensive understanding of the basics of extreme weather events to formulate effective mitigation and adaptation strategies. Analyzing the agricultural impacts of these events reveals the vulnerabilities of different crops and livestock to specific weather extremes. Additionally, disruptions in supply chains and market fluctuations amplify the challenges faced by the farmers. Protective measures, such as resilient crop varieties, improved water management, and early warning systems, are crucial for minimizing the adverse effects of extreme weather events on agriculture. The significance of proactive measures, including climate-smart agricultural practices and infrastructure development, is to enhance the resilience of farming communities.

By integrating these protective measures, stakeholders can mitigate the
impact of extreme weather events on agriculture, fortifying food systems
against the uncertainties imposed by a changing climate.

Keywords: Extreme events, Measures, Agriculture.

Introduction

Extreme weather events can be defined as rare meteorological events
that occur above a defined threshold value much above its normal
for that period. The severity of such events depends on its impact
on the natural environment as well as on human society. It is also
described as a disaster, based on "its impact causing severe damages
to physical infrastructure and leading to severe discomfort, heavy
losses to human and livestock populations within a short period".
Similarly, extreme agrometeorological events can also be defined as
the interactions between an agricultural system and extreme weather
events. One important aspect of this phenomenon is its randomness
and abruptness with which it occurs in a given region.

   In India, over 40 million hectares (12% of land) are prone to
floods and river erosion. Out of the 7,516 km long coastline, close
to about 5,700 km is prone to cyclones and tsunamis and 68% of
the cultivable area is vulnerable to drought, and a greater part of
the hilly areas of Himalayan region is at risk from landslides and
avalanches (Subbarao, 2013) The super cyclone of 1999 that hit the
eastern coast of India (Orissa State) was a major natural disaster
that affected the subcontinent in recent years. The Bengal famine in
1943 was one of the worst, recording three million deaths in India due
to severe drought and associated diseases like malnutrition. During
the last 50 years, the droughts of 1972, 1987, 2002, 2009, the heat
waves in 1995 and 1998, and the cold wave in 2003 are some of the
extreme weather events that affected the country. The worst drought
in India occurred during the last century in 1918 while the highest
rainfall all over India happened in the preceding year, 1917. However, in many scientific reports, much of the information on extreme
weather events lies scattered. Thus, it is necessary to gather information on extreme weather events and prepare a suitable database
to help scientists research on identifying appropriate adaptation

strategies to reduce their impacts on various sectors including agriculture.

The Council on Energy, Environment and Water (CEEW) in its report during the year December 2020 — through geospatial, temporal analysis — provided a detailed assessment of the impact of extreme events at a district level. It is reported that India has witnessed more than 478 weather extreme events over the last 50 years (1970–2019) and many of them occurred after 2005. Hotspots and climate change landscape at district levels for extreme events over the above period are shown in Fig. 1, providing a clear picture of climatic vulnerabilities at district level across India.

Key findings of the study are reproduced as follows:

- Three out of four Indian districts, representing a population of over 638 million people, are more prone to extreme climate events.
- Compared to the period of 1970–2005, which experienced about 250 extreme events, the period 2005–2020 recorded 310 extreme weather events and associated events. These include heat and cold waves besides extreme flood events in about 55 districts, exposing 97.51 million people annually.
- Highest flood frequency was observed in 2005, with 140 floods affecting 69 districts. In contrast, the number of affected districts increased to 151 in 2019.
- Also, a surge in disaster-prone events like heavy rainfall, hailstorms, thunderstorms, and cloudbursts landslides, was observed between 1970 and 2019.

From the weather point of view, the year is divided, in general, into four seasons, *viz.*

(i) *Winter (January–February)*: Predominantly experiencing cold waves, fog, snow storms, and avalanches.

(ii) *Hot weather season (March–May)*: Prone to heat waves, hailstorms, thunderstorms, high winds, and dust storms.

(iii) *Southwest monsoon season (June–September)*: The main rainy season over major parts of the country susceptible to tropical cyclones, tidal waves, floods, heavy rainfall and landslides, and droughts which are also prevalent in the post-monsoon phase.

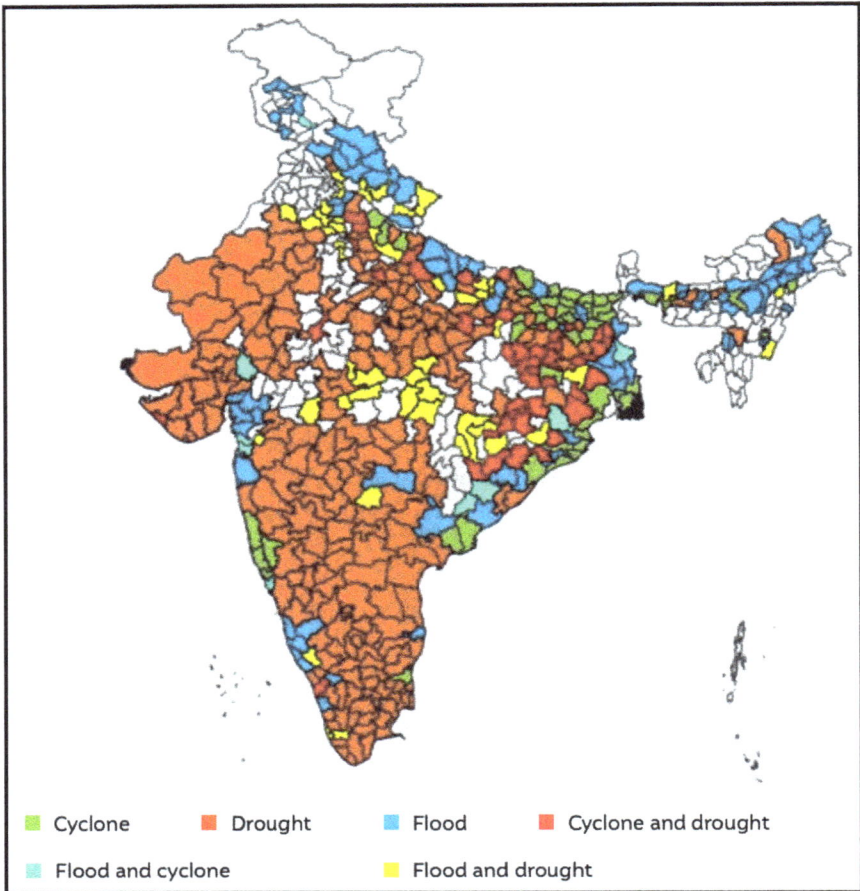

Fig. 1. Vulnerability of different districts across India with respect to cyclone, drought, flood, cyclone and drought, flood and cyclone, and flood and drought. *Source*: CEEW (2020).

(iv) Post-monsoon season (October–December) especially in the Tamil Nadu and Kerala regions.

Criteria Adopted by IMD for Identification of Extreme Weather Events

IMD (2021) adheres to the following criteria for identification of all the extreme weather events in India.

S. No	Event	Criteria
		Cold wave (hills)
		(considered when minimum temperature is below 0°C)
1	Normal	Minimum temperature is 0°C or more
2	Cold wave	Minimum temperature departure >−4.5°C to −6.4°C from its normal value
3	Severe cold wave	Minimum temperature departure >−6.5°C or more from its normal value
		Cold wave (plains)
		(considered when minimum temperature is below 10°C)
4	Normal	Minimum temperature is 10°C or more
5	Cold wave	Minimum temperature 4.5°C to 6.4°C below normal
6	Severe cold wave	Minimum temperature 6.5°C or more below normal
		Drought
7	Moderate	When SWM rainfall departure is between −26% and −50%
8	Severe	When SWM rainfall departure is less than −50
		Dust storm
9	Moderate	When surface wind speed is 41–61 km/h (in gusts) and visibility is less than 1,000 m but more than 500 m
10	Severe	When wind speed is 62–87 km/h (in gusts) and visibility is less than 200 m
		Hailstorm
11	Slight	Thunderstorm with hail sparse and small in size and often mixed with rain
12	Moderate	Thunderstorm with hail fall abundant enough to cover the ground mixed with rain
13	Heavy	Thunderstorm with at least a proportion of large hail stones mixed with rain

(Continued)

S. No	Event	Criteria
	Heatwave	
	(considered when maximum temperature of a station is above 40°C in plains and above 30°C in hill regions)	
14	Moderate	Maximum temperature departure is 4.5°C to 6.4°C above normal value
15	Severe	Maximum temperature departure is ≥6.5°C above normal value
	Snow	
16	Heavy	64.5 mm to 115.5 mm in 24 h
17	Very heavy	115.6 mm to 204.5 mm in 24 h
18	Extremely heavy	205.5 mm and above in 24 h
	Thunderstorms	
19	Moderate	Thunderstorms with maximum surface wind speeds of 41–61 kmph (In gusts/squall)
20	Moderate to heavy	Thunderstorms with maximum surface wind speeds of 62–87 kmph (In gusts/squall)
21	Severe	Continuous thunder and lightning, heavy rains and Maximum surface wind speeds ≥88 kmph

Extreme Events and Its Impacts on Agricultural Production

The mean climate of a region regulates agricultural crops and its production. Any large-scale deviation of weather due to extreme events exerts a negative influence on crop production. However, in some cases it can also be positive.

Positive effect on agriculture

Though there are many negative impacts of extreme events such as droughts, tropical cyclones, and floods on agricultural outputs, there are also several positive impacts or benefits of extreme events such as fixing of atmospheric nitrogen due to thunderstorm activity, maintenance of the fertility of the basin soils due to river flooding, etc. Droughts/dry seasons are especially essential during the pre-harvest stage, which helps to record the highest crop production. The incidence of pests and diseases is also observed to be low during periods of drought.

Negative effects on agriculture

The negative impacts of extreme weather events can either be direct or indirect, which can lead to damages or losses to agricultural crops. Direct impacts are due to physical contact of the events with the production system, people, animals, and their property. Indirect impacts of extreme events are those induced by events such as food shortages and escalation of prices of all essential commodities. Direct and indirect effects of weather events as discussed by Motha (2011) are briefly presented as follows:

(a) *Rainfall*: It causes direct damage to plant parts like branches, flowers, ear heads, etc. Soil and water erosion from cultivated fields, fertilizer losses, water logging in the standing crops; occurrence of droughts and floods, landslides, and impeded drying of produce are a few examples. Also, these conditions will be favorable to crop pest/disease development and will have a negative effect on pollination and on pollinators.
(b) *Wind*: It leads to physical damage to all plant organs, uprooting of whole plants, lodging in crop plants topsoil erosion, and increased irrigation demand due to excessive evaporation. Wind also aggravates bush or forest fires.
(c) *High temperature*: It brings about increased thermal stress, higher evapotranspiration rates, induced sterility in certain crops, reduced crop growth duration, and create favorable conditions for the survival of pests and their damages to crops during winter. Higher temperatures at night increase respiration losses.
(d) *Low temperature*: It damages cell structure (frost) and reduces crop growth, particularly during prolonged cold wave conditions.
(e) *High cloudiness*: It is responsible for lower ET rates, increased incidence of diseases, and poor growth.

(f) *Hail impact*: It does significant damage to all crops and orchards particularly at critical phenological stages and to infrastructure facilities.
(g) *Lightning*: It causes damage to infrastructure facilities and loss to humans and livestock and initiates wildfire.
(h) *Snow*: Heavy snowfall damages the reproductive organs of plants and vegetable crops.

Strategies Adopted in Areas with High Weather Risk

The following are the agricultural strategies adopted to reduce the disastrous potential by practicing appropriate strategies and timely preventive measures developed by various research organizations across the globe, as reported by Rao (2013).

Techniques in drought management

In arid, semi-arid, and marginal areas, the probability of drought incidence is at least once in 10 years. Therefore, it is important to develop viable land-use plans including agricultural crop plans, and suitable weather-based agro-advisories during main crop season. Drought is the result of the interaction of land use by the human population and the rainfall regimes. Therefore, there is an urgent need for a detailed analysis of rainfall data of these regions to develop methods for predicting rainfall events in advance (long-range forecast). More emphasis should be given to various drought management practices with reference to rainfall variability and the availability of other resources including human resources. Agricultural planning and practices need to be worked out with respect to total water requirements for each individual agro-climatic zone. Short-duration crops with relatively low water requirements need to be encouraged in drought-prone areas. Water harvesting potentials need to be computed and structures to be designed and developed to store excess runoff. Irrigation, through canals and groundwater resources, needs to be monitored for optimum utilization of water resources to avoid soil salinity and excessive evaporation losses.

A food reserve to meet the requirements of both humans and livestock for up to two consecutive drought years needs to be established. Strategies must be developed to minimize the impact of drought on crop production. Recurring droughts can be avoided through varietal manipulation and its effects can be minimized by adopting drought-resistant varieties which are drought-escaping or drought-resistant at different growth stages. For midseason drought management, corrective measures such as reducing plant population and suitable fertilization or weed management can be adopted. In high rainfall regions, excess rainfall during wet spells can be harvested in farm ponds or in village tanks and can be used as lifesaving irrigation during a prolonged dry spell. Most of the strategies that developed over the years are location, time, crop, crop stage, and socioeconomic condition-specific. Developing a combination of such strategies shall help make agriculture sustainable and profitable. Participation of stakeholders at the community level will help manage droughts effectively.

Preparedness for cyclone management in agriculture system

Disaster preparedness for cyclones consists of an action plan needed to reduce the loss of human lives and damage to infrastructure and agricultural crops. Preparedness of the agriculture system for cyclones includes timely harvesting of crops if matured, transport and safe storage of the produce, etc. Frequent checking of irrigation canals and embankments of rivers in the risk zone should be strengthened and repaired immediately to avoid breaching and flooding into the surrounding areas. Safety places for humans and livestock should be constructed on priority in all the coastal regions to avoid storm surges and gale winds during cyclones. As a long-term strategy, the farmers in the areas prone to cyclones are advised to take up the cultivation of short-duration variety crops which are not easy grain shredders. Further, rice varieties like Swarna-Sub are submergence-resistant and good for flooding areas.

Measures to manage floods, heavy rainfall damages in the agricultural sector

- Soils that have been saturated before an extreme weather event are likely to be more affected by flood than soils that are relatively dry. Recently tilled fields that are devoid of vegetation are more prone to soil erosion.

- Vegetation can act as a physical barrier to moving water, which can reduce flood severity and impacts.
- Water storage systems such as reservoirs and dams, which can hold most of the incoming water, can effectively reduce flood damage.
- Long-term weather data analysis for any specific region shall help identify rainfall patterns during the growing season to estimate the water availability to raise crops under dryland conditions. Also, it shall help work out the water harvesting potentials to provide supplemental irrigation.
- Given the increased climate variability associated with climate change, it is essential to analyze long-term weather data.
- Rice crops can grow effectively in saturated and even under submerged conditions and are more appropriate for locations that are flood-prone.
- Other crops are unsuitable for such conditions and would not be promoted as alternatives to rice crops.
- Floodplain maps using GIS with appropriate information on the probabilities of certain amounts of precipitation, area, and depth of flooding water should be developed and used in risk assessments and agricultural planning.
- Improved drainage system with balanced environmental considerations such as wetland protection and downstream residents must be constructed to avoid water logging and flood damages.
- All the community members' participation is another important measure in managing the available water in watersheds more efficiently.
- Promote only ecologically appropriate policies for human settlements and agriculture in the flood plains.

Measures to manage high winds in agricultural sectors

- Damage to crops by strong winds can be minimized or prevented by using windbreaks/shelterbelts across the prevailing wind direction which are either natural (e.g., trees, shrubs, or hedges) or artificial (e.g., walls, fences) barriers.
- Well-designed shelterbelts are effective in stabilizing agricultural production in regions where strong winds can impose severe moisture stress on standing crops.

- Windbreaks also reduce soil erosion from the soil surface and improve the moisture availability to crop plants for slightly longer periods. Reduction in wind speed on the leeward side of shelterbelts reduces crop ET.
- Select crop varieties which are resistant to lodging may escape the impacts of strong winds during the sensitive crop growth stage. Also tying plants into bunches, can provide enough extra strength to the plant community and provide sufficient space for the free flow of wind through the plant canopy thereby reducing the damaging impacts of wind on crops.
- Improved crop management practices like crop rotations and intercropping can also provide adequate cover to plant communities against strong winds.

Strategies in crop management against dust storm/sand storm

- Afforestation in arid regions is one of the measures to protect the topsoil erosion from dust storms.
- Soil resistance to erosion can be accomplished by carefully selecting proper cultivation methods, applying mineral and organic fertilizers, sowing grass on the borders, and spraying the soil surface with various substances that enhance soil structure.
- One of the major protection strategies is to develop plant cover (in a checkered pattern) with suitable grass species before the onset of dust storm periods (e.g., sand dune stabilization). This will reduce the wind speed in the layer adjacent to the ground by forming an effective buffer zone.
- Reduction in the wind speed at the soil surface and improving the hooking of soil particles which are crucial can be achieved by establishing tree shelterbelts and windbreaks.
- Leaving stubble in fields, cultivation with non-mold board plow, spraying of chemical substances that promote the binding of surface soil particles, mulching, growing perennial grasses, intercropping, and seeding of annual crops are also important techniques that control soil loss from the cultivated lands.
- In regions prone to severe wind erosion, especially on slopes or on light soils, strip farming cultivation may be suitable.

Protection of crops from low temperature and frost damages

Frost causes considerable damage by freezing the water in the plant cells, thus killing the standing crops in temperate and subtropical climates. To minimize the damage from frost and to prepare crop plans proper knowledge on the characteristics and probable period of occurrence of the frost is essential. The methods of frost protection can be classified into passive and active forms. Passive protection method consists of site and variety selection and several cultural practices that include brushing and soil surface preparation. Additional expenditure is not required to practice this method. Active protection method replaces radiant energy losses by practicing methods such as irrigation, heating the air, smoke generation, and wind machines. These methods require additional funds and outside energy to operate the system.

Several protection measures against frost such as site selection, crop management methods, etc., are briefly described as follows:

(a) *Site selection criteria*: The site for growing crops should be selected by taking into account the prevailing climatic conditions, its slope, and the soil characteristics in that location. Low-lying areas in the field with cold air build-up are not the best locations for planting orchards and frost-sensitive crops. Removal or thinning of trees that create cold air dams is desirable. If the site has a good passage for cold air drainage, then it is considered a good site as far as frost damage is concerned. Fruit trees are usually planted on hillside slopes as it will be usually 2–4°C warmer during radiational frost conditions due to the rapid downward movement of cold air to the bottom of the valley.

(b) *Frost-tolerant cultivars*: Growing frost-resistant cultivars and varieties is another method to avoid frost damage in orchards and in field crops. Oats are comparatively more tolerant to frost than barley. However, barley is slightly more tolerant than wheat crops. The popular frost-resistant varieties have been developed with the incorporation of genes that are tolerant to freezing. Growers may refer to the available literature on the varieties that can withstand low temperatures.

(c) *Optimization of sowing dates*: The best cost-effective strategy to save field crops from frost is the choice of optimum sowing date such that the most sensitive stages like flowering and grain-forming stage escape the heavy frost periods thus reducing the impacts of frost on crop growth and on the production.

(d) *Increase heat content in the soil*: Frost frequency is higher in orchards where the soil is cultivated and covered with weeds or mulch when compared to orchards in which the soil is moist and weed-free. This is because wet soil with weed-free and compact conditions stores more heat in the daytime than dry soil covered with shade. During night time stored heat is released to the lower layers of the air in the crop plants and fruit trees, thus, minimizing the damage from frost. A dry cultivated field increases the frequency of frost occurrence because of the presence of air pockets in the soil that hampers the flow of heat to the soil thus, lowering heat storage during the day hours. Hence, keeping the soil wet with frequent irrigations by sprinklers, weed-free fields, and making it compact with rollers are some of the best techniques to reduce frost damage in orchards, vineyards, and wide-row crops.

(e) *Role of plant cover*: Growing large canopy trees within orchards provides freeze protection to some extent. For example, date palm trees in California and pine trees in southern Alabama are being used as canopy cover for citrus plantings (Perry and Bradley, 1994). "Brushing" is commonly used for protecting vegetable crops from frost damage (Samra *et al.*, 2002).

(f) *Role of nutrition*: Some of the deciduous fruit plants like peach, which is not nutritionally sound to nitrogen, are more susceptible to frost. Fruit buds of these trees are less healthy and are easily damaged by frost. In order to withstand frost in these plants, the application of nitrogen in midsummer or in the post-harvest season is recommended to induce vigor for strong bud development and delay in flowering. However, tree fruits, such as apples and pears, do not normally require mid- to late-summer fertilization due to their low fertility requirements. However, blueberries are likely to benefit from such fertilizer applications.

(g) *Role of chemicals*: Spraying of some chemicals such as cryoprotectants, anti-transpirants that could change the freezing point of plant tissues and growth regulators such as ethylene-releasing compound ethephon that could increase the cold hardiness of the buds and flowers and delayed flowering are found to be beneficial against frost damage.

(h) *Role of irrigation*: Irrigation using sprinklers is the most effective method of protection from frost as sprinkling the water in the canopy releases the latent heat of fusion when water turns from liquid to ice. As this process of formation of ice is continuous, latent heat released by water compensates for the heat lost from the crop canopy to the surrounding environment. Once irrigation starts in the night hours, it must continue up to the morning hours or till the temperatures rise to tolerable limits for the plants.

(i) *Role of heaters*: Heating in the orchards by burning waste wood material to protect from frost has been in practice for many years. Over the time period people have moved to other methods such as burning of fossil fuels which can be placed as free-standing units or connected by a pipeline network throughout the crop area. The major advantage of this system of connecting heaters is its ability to control the rate of burning and closing all heaters from the central pumping platform by simply adjusting the pump pressure. Though the initial installation costs are lower than those of other systems, the expensive fuels required to operate the system increases its operating costs.

(j) *Role of wind machines*: The main purpose of using wind machines is to circulate the warm air down to the crop level. These machines are only effective under radiation frost conditions. Wind machines that are used along with heaters provide the best protection against frost. They also have excellent advantages in frost protection by minimizing labor requirements and reducing the refueling and storage of heating supplies that ultimately end up at a low operational cost per hectare.

Strategies to prevent hail storm damages

Thunderstorms containing hail stones are formed when hot air rise from the soil surface in plumes due to high surface temperatures that cool each greater heights in the atmosphere and the moisture in the air condenses to form clouds. Due to the intense circulation pattern within the cloud, the condensed water in the thunderstorms becomes super cold. Once the super cold water comes in contact with any particulate matter in the cloud such as dust, ice crystals are formed and grow as hailstones due to collision and coalescence process within the cloud. Once these stones attain a particular weight, they reach the ground along with the rain due to gravity. The overall effect of hail damage will depend on the size of the hail, the duration of the hailstorm, and the type and growth stage of crops.

Hail storms that occur during pre-monsoon season inflict severe damage to orchards and summer crops. Their activity during monsoon season is low in the country. However, the hail storms associated with western disturbances during *rabi* season in the northern plains and hilly regions affect the crops at the maturity stage.

Netting to protect crops from hail damage, bird predation or insects is one of the best possible solutions for reducing crop damage or production loss. Using crop protection netting will save crop damage and the costs for its installation and maintenance can be recovered in the first few years of use. With proper care and handling or installation practices, the life span of the system is expected to be about 10 years. The modern system, which was developed in Europe, with integrated zipper facilitates quickly opens and closes the nets and approximately takes about 25–35 min to protect one hectare of land (Chander, 2018). This leads to savings in time and costs for fruit growers in the country. Tree shelterbelts also provide limited protection from hail damage to the crops raised under the shelterbelts. Some useful crop-specific post-hail damage strategies suggested by Bal *et al.* (2014) are reproduced in this section.

Forest fire prevention measures

Forest fires observed in India are due to manmade interferences with forest ecosystems and also due to strong summer heat which can initiate a small fire that can quickly spread to nearby plant populations due to the abundant availability of dried plant litter all around the forest floor. In addition to human influences, sometimes lightning also accounts for the sudden start of localized fires. Analysis of climatic data helps us identify the probabilities of such periods and prepare long-term plans to prevent such happenings in the future. Based on the weather–fire relationship, fire danger rating systems can be prepared for all the forest fire zones as a guide to all fire management departments in their day-to-day activities. Such a system can be of help to:

- Improve forest fire predictions and warning capability.
- Develop insurance and disaster support plans.
- Continue fire suppression efforts with the implementation of the Wildfire Management Strategy (NDMA, 2020).

Conclusions

Given the increased frequencies of extreme weather events due to global warming conditions, there is a need to improve the quality of existing (WRF) weather forecasts and alerts to a scale of at least $3\,km^2$ grid level for the benefit of all stakeholders. Applications of remote sensing technologies through satellites for identification and nowcasting of extreme elements should be encouraged. Similarly, the utilization of drone services in rescue and rehabilitation operations especially in inhospitable regions like hilly regions should be initiated. Just like providing subsidies toward fertilizers and drip irrigation systems, the Government may provide subsidies to the farmers cultivating crops in areas prone to low temperatures, frost, and also toward procurement of equipment necessary to minimize the impact of frost damage to the crops. Farmers should be made aware of the different technologies available to manage and minimize the adverse effects of weather events on their crops and livestock and provide needed assistance for timely action to save their crops. Crop insurance should be made mandatory for all farmers to save them from economic losses incurred during the occurrence of extreme weather events that adversely affect their agriculture and livestock.

References

Bal, S.K., Saha. S., Fand, B.B., Singh, N.P., Rane. J., and Minhas, P.S. (2014). Hail storms: Causes, damage and post-hail management in agriculture. *Technical Bulletin No. 5*, National Institute of Abiotic Stress Management, p. 44.

CEEW. (2020). *Preparing India for Extreme Climate Events: Mapping Hotspots and Response Mechanisms.* Report by Mohanty, Abinash, p. 41.

Chander, M. (2018). Hail Storm Protection Net. *Krishi Jagaran.* https://krishijagran.com > news > hail-storm-protection-.

IMD. (2021). *Standard Operation Procedure — Weather Forecasting and Warning Services.* India Meteorological Department, Ministry of Earth Sciences, Government of India, p. 331.

Motha, R.P. (2011). *The Impact of Extreme Weather Events on Agriculture in the United States.* Chapter 30: USDA-ARS/UNL Faculty, pp. 397–407.

NDMA. (2020). *Forest Fire Management: Global Best Practices.* National Disaster Management Authority, p. 30.

Perry, K. and Bradley, L. (1994). *Frost/Freeze Protection for Horticultural crops. Horticulture Information Leaflets.* North Carolina Cooperative Extension Service, p. 11.

Rao, V. U. M., Rao, A. V. M. S., Vijaya Kumar, P., Bapuji Rao, B., and Sastry, P. S. N. (2013). *Agrometeorological Aspects of Extreme Weather Events,* Central Research Institute for Dryland Agriculture, Santoshnagar, Hyderabad, 303 pp. CRIDA, April.

Samra, J.S., Singh, G., and Ramakrishna, Y.S. (2003). *Cold Wave of 2002-03 — Impact on Agriculture.* Natural Resource Management Division, Indian Council of Agricultural Research, p. 49.

Chapter 11

Microclimate Modification for Improved Agriculture Production

V.U.M Rao*, Latief Ahmad, and Abdus Sattar

Agromet, CRIDA, Santoshnagar, Hyderabad, India

*vumrao54@gmail.com

Abstract
Microclimate modification is a strategic approach for enhancing agriculture production. Microclimate, defined as the localized climate conditions within a specific area, plays a pivotal role in determining crop growth and yield. This chapter emphasizes the significance of purposeful interventions to modify microclimates to optimize agricultural productivity. Efficient microclimate modification involves strategic deployment of technologies such as shade nets, greenhouse structures, and agroforestry practices. These interventions aim to regulate factors like temperature, humidity, and sunlight exposure, creating a more favorable environment for crop cultivation. By manipulating these microclimatic conditions, farmers can extend growing seasons, protect crops from extreme weather events, and optimize resource utilization. Furthermore, this chapter discusses the potential of precision irrigation and mulching techniques in microclimate modification, contributing to water conservation and soil moisture management. The integration of technology, data analytics, and climate monitoring systems enables farmers to make informed decisions in real-time, adapting to changing microclimatic conditions. Microclimate modification emerges as a promising strategy to address the challenges posed by climate variability and enhance agriculture production. By adopting innovative technologies and sustainable practices, farmers can optimize microclimatic conditions, thereby increasing resilience and productivity in the face of a changing climate.

Keywords: Microclimate, Agriculture production, Climate change.

Introduction

The microclimate furnishes the environment in which plants and living beings thrive. As such plants themselves are efficient integrators of the environment. Microclimatological studies in the plant ecosystem are necessary for better understanding of the physical interaction between plants and the environment. The success or failure in sustainability of crop production depends on local weather and climatic conditions, and year-to-year variations within a region. One means of alleviating climatically induced stress is through modifications of microclimate which include any artificially introduced changes in the composition, behavior, or dynamics of the atmosphere near the ground so as to improve the environment in which crops grow. In other words, microclimate modification is an intended change in the soil–plant–atmosphere system, which alleviates stress or prevents damage with the aim of attaining improved yields. Modifications of microclimate are expected to bring about changes in one or more of the meteorological parameters depending upon the technique used to alleviate climatic stress. The interaction of plant community in relation to microclimate needs basic weather information like solar radiation, light intensity, soil temperature, relative humidity, temperature, wind force, etc., at the micro-level along with topography, vegetation, drainage, cultural practices, etc., in order understand the system better. In this chapter, we discuss modifications of microclimate to reduce the impact of extreme weather events (Brown and Rosenberg, 1972).

Windbreaks probably go far back in the history of pastoral and agricultural civilizations. Wind problems have been of major importance in determining the characteristics of agriculture in many regions. In the lower Rhone Valley of France, for example, almost all agricultural enterprises require some degree of protection against the force of the mistral winds. This has led to the culture of small fields protected by a dense network of windbreaks (Van Eimern, 1964).

We observe that grazing animals seek shelter from strong winds. This is a response, no doubt, to physical discomfort caused either by the chill in the cold wind, by desiccation in hot winds, or simply by the mechanical pressure on the animal. Plants, too, are subject to damage caused by excessive chills, high temperatures, desiccation, or direct mechanical injury. **Windbreaks** (any structure that reduces wind speed) and **shelterbelts** (rows of trees planted for wind protection) can by reducing these stresses be profoundly beneficial to the growth of plants in their lee.

Soil Temperature

Growing plants live in two media — in the air near the ground and in the upper layer of the soil. The soil is the only medium for most plants in their first stage of life before the seeds have germinated and emerged above ground. Soil temperature can be a limiting or critical factor for seed germination, root elongation, tuber development, decomposition of organic matter within the soil, and thus for the amount of CO_2 passing from the soil into the plant-air layer and into the atmosphere. The optimum and minimum soil temperatures for seed germination for some crops are given as follows:

Crop	Optimum soil temp. (°C)	Minimum soil temp. (°C)
Wheat	15–18	3–4
Maize	20	8–9
Soybean	25	8–9
Field beans	21	10
Potato	—	7–9
Cotton	21	—

The range of temperature for germination of a crop defines its sowing date. Soil management which results in the increase/decrease of soil temperature so as to be above the minimum but close to the optimum will be desirable. Soil temperature is also important for plants for which parts are harvested below the soil surface. For example, the tubers of potato develop best at temperatures about 18°C but their growth is reduced at temperatures above 20°C, and is stopped entirely at about 28°C. Some techniques to modify soil temperature are described in the following sections.

(a) *Soil cultivation and treatment*: Cultivation may increase the porosity of the soil and consequently improve its aeration and

increase its temperature; it may also improve drainage. Cultivation of the soil also influences agrometeorological conditions
in soil and air — evapotranspiration, air temperature near the
ground, and its daily variation. Such changes in the state of soil
are brought about by changes in surface albedo as a consequence
of soil cover, shallow mulch, and reshaping the soil surface as
ridges and furrows.

(b) *Effect of soil surface color on soil temperature*: Differences in soil
color imply differences in the coefficient of reflection of short-
wave solar radiation and hence different amounts of solar energy
absorbed by the surface. So net radiation of the soil surface can
be changed by altering surface color. This method is used in
agriculture to warm the uppermost layers of the soil or to prevent high temperatures, which are dangerous for young plants
and seedlings. For example, use of a thin layer of white pow-
dered lime on black cotton soil reduces the surface temperature
up to 15°C. Similarly, during cooler period, in temperate areas,
dark materials can be applied to increase the soil temperature
for encouraging earlier germination.

(c) *Effect of ridge-furrow system on soil temperature*: Some agronomic practices such as the sowing of crops on directional ridges
can be used to alleviate cold stress especially during early growth
stages which are usually more sensitive (see Fig. 1). Soil temperature at the surface as well as at shallow depths has been found
to be higher by several degrees on south-facing slopes of the
east-west drawn ridges or beds. This benefit is being exploited in
winter maize and sunflower cultivation to encourage early crop
establishment during the winter months in Punjab where the
seeds of the crop are sown on south-facing slope of the ridges.

Air Temperature

The temperature at the top of the canopy can be increased or
decreased by changing the radiative fluxes at the top of the crop.

(a) *Burning of debris*: Burning of debris is used to create smoke
in calm conditions to reduce long-wave radiation losses at night
and help prevent frost damage. This technique is used in fruit
orchards when a frost could harm sensitive buds.

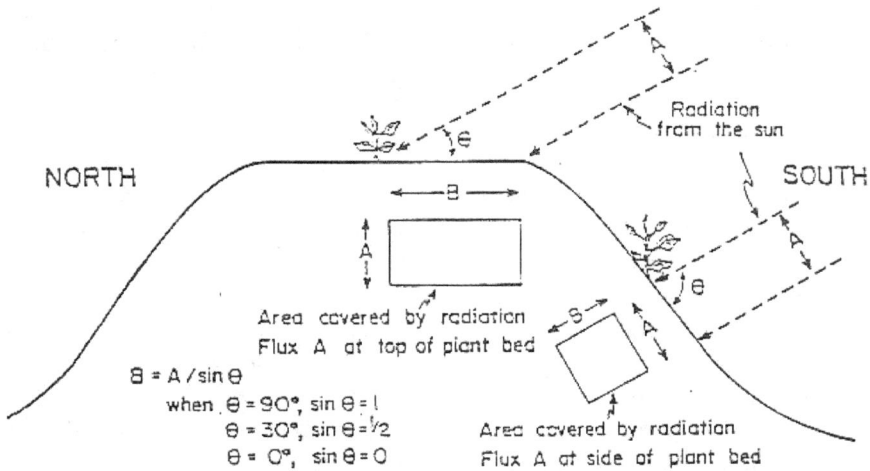

Fig. 1. Amount of radiation at the top and side of bed oriented east-west, in the northern hemisphere, where θ is the angle of the rays to the surface of the bed.

Source: Yamaguchi (1983).

(b) *Porous crop covers*: These are often used to reduce incoming radiation and the radiant heat load by shading the surface. These covers also reduce the transfer of sensible and latent heat from the crop which tends to bring about an increase in temperature in the crop. Various plant protective structures and methods used to alter radiative fluxes from the crop include:

   (i) *Hot caps*: Plastic material acting as a miniature greenhouse is used to cover one or more plants in special structures. Cold and sensitive plants such as tomatoes and floricultural plants are individually protected with plastic bags during the winter period.

   (ii) *Cold frames*: These are made of wood, metal, or concrete with plastic or glass-covered tops. These structures shelter the plant from cold winds, making the air several degrees warmer than the outside. Cloth or straw mats are often used to cover the glass to reduce heat losses at night.

   (iii) *Plastic tunnels*: Plastic sheets are supported over plant rows by wood or bamboo sticks. These structures act as miniature greenhouses.

(c) *Greenhouses*: These are made of glass or plastic and the structure is usually large enough to work inside. The greenhouse influences the plant environment in a number of ways:

   (i) The exchange of both sensible and latent heat by convection is restricted, resulting in an increase in the plant temperature.

  (ii) Due to increased atmospheric humidity inside the greenhouse, the plants are protected against excessive rates of water loss.

 (iii) Glass allows penetration of incoming short-wave radiation during the day but restricts outgoing long-wave radiation, thus increasing the temperature inside the greenhouse.

 (iv) Furthermore, the long wave radiation from the sky is absorbed by the glass surface and re-emitted. The plants below receive a greater amount of long-wave radiation from the glass surface than they would from the sky. This raises the plant temperature and helps provide protection from radiative frost.

(d) *Baskets and shade boards*: These structures are used to protect the transplants from excessive solar heating during the summer in hot climates.

(e) *Wind machines*: Wind machines are used to alter sensible heat fluxes at the top of crop surface at night time. These machines mix the warmer air above with colder air near the crop surface. This technique works only when an inversion exists in which the air above is warmer than the air near the crop surface. Temperature inversions are common at night under clear skies. Wind machines have been used in California to protect citrus orchards.

(f) *Use of shelters*: The shelterbelt, or windbreak, is a barrier consisting of a line of trees, bushes, hedges, soil embankments, stonewalls, or fences. Its objective is to reduce the horizontal wind speed near the ground in regions where strong winds are a problem (Fig. 2). Wind shelters decrease the velocity of the wind by absorbing some of its momentum. The effectiveness of a windbreak depends on its (i) height, (ii) porosity, and (iii) length. The higher the windbreak, the farther downwind will be its effect. The length of the sheltered zone is usually expressed in multiples

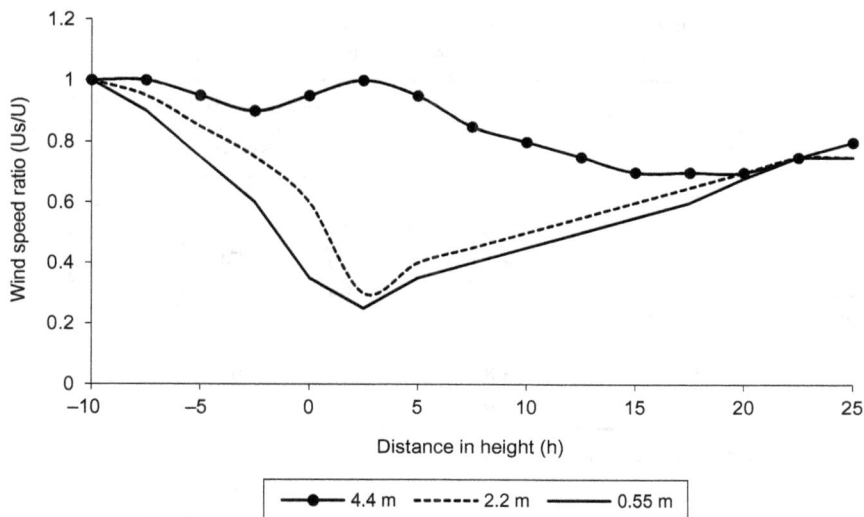

Fig. 2. Influence of a dense windbreak on the ratio of wind speed in shelter (Us) and in the open (U) (Oke, 1987).

of the height (h) of the barrier (see Fig. 2). A dense barrier protects an area of about 10–15 h downward. When the porosity of the shelter increases to about 50%, the downwind influence approaches 20–25 h.

The shelterbelt affects the radiation balance. Solar and net radiations are reduced significantly in the areas shaded by windbreaks. In the northern hemisphere, the east-west oriented windbreaks are more effective than the north-south oriented windbreaks. Areas to the north of an east-west windbreak will be shaded for a longer time while to the south there will be increased irradiance throughout the day due to reflection from the barrier.

The shelterbelt affects air temperature and humidity. The data in Table 1 show a comparison of mean daytime air temperature and relative humidity at 0.5 m height in a sugarbeet field on several days in the open and sheltered plots with corn rows serving as shelterbelts. Generally, the daytime temperatures are greater in sheltered fields than in open fields due to the reduction of wind speed and consequent reduction of mechanical turbulent mixing. At night, the reduction of windiness and turbulence

Table 1. Mean daytime (0600–1800) air temperature and relative humidity at 0.5 m height in open plots and in plots of sugarbeet sheltered by corn rows.

Date	Air temperature (°C)			Relative humidity (%)		
	Open	Shelter	Difference	Open	Shelter	Difference
Aug 10	22.2	22.5	+0.3	77	92	+15
14	19.5	19.1	-0.4	54	60	+6
16	25.2	26.8	+1.6	76	74	−2
18	24.6	28.1	+3.5	66	69	+3
25	23.1	24.6	+1.5	73	74	+1
Sep 1	20.2	22.6	+2.4	86	88	+2
3	22.7	25.7	+3.0	57	65	+8
5	18.4	18.5	+0.1	59	64	+5

Source: Brown and Rosenberg (1972).

in the sheltered fields causes temperature inversions to inten-sify. Therefore, the air is often colder at night in sheltered fields than in open fields with the incidence of dew and frost thereby increased. In the sheltered field, the amplitude of diurnal tem-perature changes is also higher which can influence the sheltered crops.

(g) *Mulching*: Mulching is the application, or creation, of some soil cover, which reduces the vertical transfer of heat and water vapor. Mulching may consist of the following:

(i) A loose layer of topsoil (ploughing/hoeing produces a mulch);
(ii) Cut or gathered vegetation material such as grass, weeds, straw, tree leaves, etc.;
(iii) Crop residues or stubble;
(iv) Manufactured materials such as paper and plastic.

Mulching can have a marked effect on the soil environment. Mulching of soil surface

(i) reduces evaporation due decrease in the availability of vapor-ization energy below the mulch. Most mulches are an effective barrier to the upward flow of water vapor.
(ii) Reduces diurnal amplitude of soil temperature.
(iii) Indirectly increases moisture content of the soil.
(iv) Raises heat capacity of the soil.

In the summers, the aim of mulch is to conserve soil water while in winters it is used to conserve soil heat and prevent frost damage. Different kinds of mulches and their effects are described as follows:

(1) *Pulverized soil*: The simple, shallow cultivation of a soil surface creates a loose mulch of dry soil which reduces the thermal conductivity of the soil, because the air between soil particles acts as insulation. The surface temperature of such loose soil is likely to be higher than a soil surface, which is left undisturbed. The loose topsoil layer also decreases evaporation because of the rupture of soil capillary connections, inhibiting an upward movement of soil water from deeper layers.

(2) *Straw mulch*: Straw or similar materials have a greater insulating effect than an equivalent depth of pulverized soil. On hot days, soil temperature under straw can be lower by several degrees than without mulch. Straw mulch also helps to insulate the soil on cold days.

(3) *Plastic mulch*: Clear or colored plastic can be used to provide insulation. Clear plastic warms the soil more than black plastic. Plastic also prevents the loss of moisture from the soil. Black plastic mulch is also effective in controlling weeds.

(4) *Asphalt or petroleum mulch*: Water emulsion of asphalt has been shown to be effective as clear polyethylene in insulating the soil. However, asphalt mulch is practical only when petroleum is inexpensive. A comparison of soil temperature under different mulch materials is shown in Table 2.

Table 2. Effect of asphalt and plastic mulches on soil temperature.

| Depth (cm) | *Average temperature (°C) (11 am to 3 pm) | | | |
	Non mulched	Asphalt	Clear poly	Black poly
0	27.2	31.1	31.1	27.8
2.0	25.0	28.3	28.9	26.7
7.5	19.4	21.7	21.1	20.0
15.0	15.0	16.7	16.1	15.6
30.0	13.0	14.4	13.9	13.9

Note: *Air temperature = 23.3°C; Width of mulch = 30 cm.
Source: Takatori *et al.* (1964).

Irrigation and Moisture

(a) *Frost protection by flood irrigation*: Protection from frost can be provided prior to frost incidence. The flood water and moist soil keep the soil warm during periods of freeze damage by increasing the heat capacity of the soil; increasing heat conduction, and releasing some latent heat if freezing occurs. This is one of the practical options for the farmers to adopt when frost incidence is forecasted by the weather man. The soil should be kept moist by frequent but light irrigations during expected periods of frost occurrence.

(b) *Sprinkler irrigation for heat stress reduction*: Sprinkler irrigation which wets the crop surface helps reduce the heat stress. The application of water to the leaf surface decreases the leaf temperature by evaporative cooling of the applied water. Sprinkling for heat stress reduction can increase crop quality and yields. (Cheness *et al.*, 1979).

(c) *Sprinkler irrigation for frost and cold protection*: Cold protection is achieved by overhead sprinkling of water, which then freezes on contact with the crop, with the accompanying release of heat of fusion from the freezing ice-water film. If enough water is supplied to maintain the ice-water film, the plant temperature will remain near 0°C. If the ice-water film is not maintained then the temperature can fall below 0°C with resulting damage to the crop. Sprinklers are also useful in providing cold protection to fruit buds in orchards. The common practice is to use sprinklers to supply heat (from the heat of fusion) to the orchard for protection during freezing conditions particularly when the buds have grown to a vulnerable stage of development (Griffin & Richardson, 1979).

References

Brown, K.W. and Rosenberg, N.J. (1972). Shelter effects on microclimate, growth and water use by irrigated sugarbeets in the great plains. *Agricultural Meteorology*, 9, 241–263.

Cheness, J.L., Harper, L.A., and Howell, T.A. (1979). Sprinkling for heat stress reduction. In B.J. Barfield and J.F. Garber (Eds.), *Modification of the Aerial Environment of Plants*. American Society of Agricultural and Biological Engineers, St. Joseph.

Griffin, R.E. and Richardson, E.A. (1979). Sprinkler for microclimate cooling and bud development. In B.J. Barfield and J.F. Garber (Eds.), *Modification of the Aerial Environment of Plants*. American Society of Agricultural and Biological Engineers, St. Joseph.

Oke, T.R. (1987). *Boundary Layer Climates*. (2nd ed.). Methuen, London.

Rosenberg, N.J. (1974). *Microclimate: The Biological Environment*. John Wiley, p. 315.

Takatori, F.G., Lippert, L.F., and Whiting, F.L. (1964). The effect of petroleum mulch and polyethylene films on soil temperature and plant growth. *Proceedings of the American Society for Horticultural Science*, 85, 532.

Van Eimern, J., Razumova, L.A., and Robertson, G.W. (1964). Windbreaks and shelters. *WMO Tech. Note No. 59*, p. 188.

Yamaguchi, M. (1983). *World Vegetables*. AVI Publishing Company Inc., Westport.

Chapter 12

Micro-Climate and Its Management for Building Resilience

M.K. Nanda

Department of Agricultural Meteorology & Physics, Bidhan Chandra Krishi Viswavidyalaya, West Bengal, India

mknanda@bckv.edu.in

Abstract

Microclimate management is building resilience within agricultural systems. Microclimate, the localized atmospheric conditions within specific regions, significantly influences crop performance and overall agricultural productivity. As climate change exacerbates weather variability, understanding and effectively managing microclimates become imperative for enhancing resilience in agriculture. The utilization of various strategies for microclimate management includes agroforestry practices, precision agriculture technologies, and green infrastructure. These approaches aim to regulate temperature, humidity, and other microclimatic factors, mitigating the adverse impacts of extreme weather events and promoting stable and optimal conditions for crop growth. It integrates data-driven decision-making processes, incorporating real-time weather monitoring, and predictive modeling. This allows farmers to proactively respond to changing microclimatic conditions, optimizing resource allocation and enhancing adaptive capacity. The significance of microclimate management extends beyond immediate crop benefits; it contributes to broader agroecosystem resilience by promoting biodiversity, soil health, and sustainable water management practices. Ultimately, by focusing on microclimate management, this chapter underscores a holistic approach to building resilience in agriculture, offering a pathway toward sustainable and climate-resilient food systems.

Keywords: Microclimate, Resilience, Management.

What is Microclimate?

Crop microclimate refers to the climate just above and within the crop canopy and in the soil root zone that can be influenced by day-to-day management practices at various time scales (Stigter, 1994a). Microclimate consists of a myriad of climatic conditions that come together in localized areas on the Earth's surface (Chen *et al.*, 1999). Microclimates derive their special characteristics from the surface features, topography, as well as land use and land cover. Microclimate is the determining factor of all agroecological processes like plant regeneration and growth, soil respiration, nutrient cycling, and pest and pathogen dynamics.

The crop zone contains a number of microclimates, sometimes grouped under terms such as "crop climate" or "ecoclimate". Management efforts by traditional farmers are aimed at modifying one or more of the microclimates in the crop zone (Wilken, 1972). The best crop microclimate is one that provides the most favorable environment for the desired plant response, that is, the response that maximizes crop productivity.

Microclimate modification is an attempt to change or regulate the elements of climate on a micro scale, resulting in a climate that is favorable for plant growth. Temperature, humidity, wind and turbulence, dew, frost, heat balance, and evaporation all influence microclimatic conditions. Key plant responses to microclimate can be managed for either radiation budgets, heat balances, or moisture balances (Stigter, 1994b).

The objective of this discussion is to develop an understanding of the interaction between vegetation and microclimate, and to provide practical guidelines for microclimate management interventions toward sustainable crop production under changing climatic scenarios.

Importance of Microclimate Modification under Changing Climatic Scenario

Adaptation and mitigation are the important features of all climate change debates. But the real challenge is to create production systems that can withstand extreme weather events like rising temperature, untimely heavy rainfall, waterlogging, stormy winds, frost, etc., and to ensure sustainable production under the changing climate scenario. There is little chance that humans in the near future will be able to modify the climate, most notably temperatures, on any large scale (Gliessman, 2015). Meanwhile, microclimates go largely unobserved and unattended. This is a huge missed opportunity. When zooming in on landscapes and the agro-ecological systems within them, there is much that can be done, to the extent that a large share of the effects of global climate change can be buffered by building the microclimatic resilience of the landscape. Micro-climate management offers much potential as a third way next to adaptation and mitigation that builds ecosystem resilience and brings a positive impact on agricultural systems and biodiversity.

How Microclimate Differs from Routine Climate Observations

The routine weather observations involve measurement of temperature, humidity wind speed, etc., in the space of more than one meter above the ground and away from structures and tall vegetation to avoid local influences. However, most crop plants spend their entire lives or at least the critical stages of early growth well below this level. Conditions in this zone are markedly different from those even a few meters aloft. The solar radiation is reflected or absorbed at the crop surface. Heat and moisture exchanges with the atmosphere occur within the crop zone. Furthermore, the vegetated areas are three-dimensional structures, receiving radiant energy on leaf, stem, and soil surfaces, and reflecting and radiating energy throughout the soil-plant structure. Microclimate is characterized by steep temperature and moisture gradients because of relatively slow fluxes by molecular conduction and diffusion in the soil and in the lowest (laminar) layer of the atmosphere and reduced wind speed near the ground.

Biophysical Processes That Define Microclimate

The energy and water exchange processes at the surface–air interface play a critical role in defining the microclimate of a typical surface. The energy exchange process involves all three modes of transfer

namely radiation, convection, and conduction. The net energy gain
or loss by a surface object may be explained by the following radiation
balance equation:

$$R_N = R_S{\downarrow} - R_S{\uparrow} + R_L{\downarrow} - R_L{\uparrow},$$

where R = Radiation, S for short wave, and L for long-wave radiation. With reference to the surface the $\uparrow$ indicates upward direction
(+ve) and the $\downarrow$ represents downward direction (−ve). The resultant
energy of these four components resultant of the four components,
i.e., net radiation (R_N) represents the energy gain or loss by the
surface.

The net radiation is disposed of as sensible heat flux, latent heat
flux, ground heat flux, and net storage of energy following the energy
balance equation.

$$R_N = S + L + G \pm \Delta H,$$

where, S = Sensible heat flux that represents the convective heat
transfer through turbulent flow or laminar diffusion, L = Latent
heat flux through evaporation or evapotranspiration, G = Ground
heat flux, i.e., conduction of heat across the soil layers and $\pm \Delta H$ represents heat storage (+ve) or release (−ve) by the system, i.e., heating or cooling of the system and also the energy used for metabolic
processes like photosynthesis. Unlike the radiative flux, the non-
radiative flux is +ve when the direction of energy flow is outward
from the surface and negative when the energy is received by the
system.

The formation of microclimate around a typical land cover surface, i.e., barren land, water body, crop field, or wet marshy land or
settlement area is mainly driven by the energy and water exchange
processes. As the water exchange process involves a significant
amount of energy transfer through latent heat the use of water in
bringing quick change in the microclimate has been recognized since
ancient times. Any microclimatic modification thus can be achieved
through the manipulation of energy and water dynamics of the system. The agrarian community around the world has been following a
series of practices consciously or unconsciously designed to suitably
modify the drivers of energy and water exchange processes as per
the specific requirement for crop cultivation in the specific region.
The present discussion will focus on the scientific basis of different

microclimate management practices and their potential adaptation for mitigating the adverse impact of climate change.

All microclimate management or manipulation practices are designed to meet the following objectives:

(a) Changing water relation;
(b) Changing radiative transfer;
(c) Changing aerodynamic properties of the surface;
(d) Changing thermal environment.

Changing water relation

Irrigation is the most common practice that causes a quick change in the crop microclimate in various ways. Water has higher specific heat, higher thermal conductivity, as well as lower albedo as compared to dry soil. Besides that, the evaporation from water in moist soil causes loss of energy through latent heat of vaporization. When irrigation is given to a crop stand the energy balance is changed immediately due to the following reasons:

(a) Albedo becomes low which causes more absorption and storage of solar radiation.
(b) Latent heat flux increases causing loss of heat and consequent cooling of the surface.
(c) Decreasing soil/crop surface temperature (due to evapotranspiration) causes a decrease in outgoing long-wave radiation.
(d) Application of water increases the specific heat of the soil and thus the heat storage capacity of the soil is enhanced.
(e) Water application increases the conductivity of soil and as a result, the heat is redistributed in the soil matrix. The surface heat is carried to the deeper layer more efficiently.

As a result of changes in thermal properties the soil/crop surface temperature suddenly decreases if irrigation is applied during the daytime. On the other hand, the net radiation or rate of energy absorption and storage increases. This process is judiciously applied for microclimatic modification toward protection against both heat injury and clod injury (or frost). Irrigation is given during the morning to increase the specific heat of the soil so that more energy is required for a unit increase in temperature and at the same time increased evapotranspiration causes cooling of the crop surface. Thus, the heat injury is checked to a great extent. Under extreme cold conditions irrigation during the afternoon conserves the temperature of the crop stand and protects the crop from frost.

Changing radiative transfer

Managing the amount of radiant energy that enters or leaves the soil-plant system is the most basic way to modify microclimates. A number of techniques have been followed by the traditional farmers of different regions for intensifying, reducing, or redistributing radiation within the crop environment. The basic approaches include:

(a) *Changing the surface albedo*: The albedo of surface materials is responsible for the amount of solar radiation reflected or absorbed by the crop stand. Changing the albedo alters the amount of energy available for heating or evaporation. Generally, light-colored, smooth surfaces have higher albedo than dark, rough surfaces. Albedo can be changed in a number of ways. Many of these practices are followed by the farmers of different regions of the world and in some cases changing the albedo is not the primary purpose. Some popular examples are discussed here.

- **Tillage practice** (plowing) increases surface roughness and thus reduces reflectivity, but this is hardly the primary purpose. In case of summer plowing which is normally practiced to expose the subsurface to the scorching sun and thereby killing the microorganisms, the change in albedo and roughness of the soil surface plays a significant role.
- The role of **irrigation** in reducing surface albedo has been discussed earlier.
- Application of **ash and charcoal** from charcoal-derived burned materials by the *Maori* farmers (indigenous Polynesian people of mainland New Zealand) causes a substantial increase in surface albedo and changes the thermal environment suitably for crop cultivation.
- Tibetan natives reportedly **throw rocks on the snow-covered fields** to hasten snow melting. A thin rock covering helps quick melting of snow by decreasing the albedo and

absorbing more solar radiation. Similar effect is obtained by **spreading soot and dust** (by Russians to clear snow from Siberian croplands).

- **Raising rock walls and pavements** that reflect light into the shady undersides of grapevines helps to redistribute sunlight within the crop zone.
- Use of **mulch** (colored plastic mulch or dry straw mulch) as a common practice adopted by the farmers not only restricts evaporation but also changes the soil albedo and restricts long-wave radiation loss from the soil and thereby conserving the thermal environment.

(b) *Managing the shade*: Altering the availability of radiation by artificially managing the shade is one of the oldest practices adopted since ancient times when the farmers started clearing and burning the forests to develop the croplands. Although "slash and burn" practice is never considered environment friendly, extensive farming in forested areas is impossible until forest microclimates are replaced by crop climates by converting woodlands to open fields. Different shade management practices include:

(i) The radiation distribution in crop stand is controlled by **managing the orientation** of furrows and row cropping. Plants in north-south rows receive more solar radiation than those in east-west rows. Soil heating also is increased by proper furrow alignment. This may be an effective practice in the middle and higher latitudes.

(ii) **Pruning and thinning** of canopies in the orchards regulate the microclimate by facilitating the penetration of solar radiation into the canopy.

(iii) In the tropics, **multistoried gardens, agroforestry, silvipasture, and shade trees** are popular practices to manage radiation distribution where the taller plants shade those nearer the ground and are themselves productive. Shade trees are grown to protect the young seedlings and the shade-loving plants in the main field against overexposure to sunlight. Shade trees are popular in vanilla, coffee, tea, and cacao plantations.

(iv) Presently the insolation is controlled by using **shade nets** of different colors as well as different percentages

of transparency. Such structures protect plants from the
adverse effects of wind and also help in regulating the temperature of the crop microclimate.

(v) **Covering with straw/crop residues** to protect young
seedlings is a common practice in the tropics. The type of
material used as covering material (mulch) is important for
thermal regulation of soil microclimate.

(c) *Managing inclination angle*: The angle of incidence of solar radiation influences its intensity at the receiving surface. Several measures may be taken to manipulate the angle of incidence and
extend the active surface for trapping solar radiation.

(i) In undulated plains, **judicious use of slope** may be helpful. The south-facing slope in the N hemisphere and north-
facing slope in the S hemisphere gets more radiation.

(ii) **Furrows can be used for trapping heat and reflecting
light** onto the plants. Radiation intercepted by an **angled
or vertical surface** may be reflected into the crop zone or
may be absorbed and reradiated as long-wave energy.

(iii) **Raised ridges and walls** extend the active surface
upward, making them advantageous for shrub, vine, and
low tree crops. In Great Britain and France, the raised stone
walls extend the active surface for trapping solar radiation
and create a "warmer climate" which helps in the cultivation of fruit trees, grapes, and vegetables.

Changing the aerodynamic properties of surface

Most of the energy, as well as mass (water, CO_2, and other gases)
exchange between surface and air occurs through turbulence in the
surface boundary layer, and this transfer process defines its microclimate. Hence, the aerodynamic properties of the surface form a vital
component of microclimatic modification. The aerodynamic properties can be modified by (i) shelterbelt; (ii) changing the roughness of
the surface; (iii) protective cover.

The shelterbelts protect the plants from blowing sand. Besides
solid structures (stone walls) there are numerous varieties of trees,
shrubs, and reeds that are grown as shelterbelts. The amount and

extent of wind-speed reduction is determined by the height, orientation, and permeability of shelterbelt elements. Windbreaks and shelterbelts increase aerodynamic resistance of the surface boundary layer thereby reducing both sensible and latent heat flux. Semipermeable windbreaks produce less violent eddies and are probably more effective than solid walls.

By increasing the surface roughness, the aerodynamic resistance increases. The roughness of the soil surface can be altered by managing tillage practice, proper land shaping, managing plant canopy geometry, etc. High aerodynamic resistance decreases the rate of energy and mass exchange in the surface–air interface.

Changing the thermal environment

The temperature is the most critical factor for growing any crop in a given growing environment. The thermal environment is the result of radiation balance, heat storage capacity of the system, as well as heat dissipation through turbulent transfer in the form of sensible heat and latent heat.

Various microclimatic modification measures discussed in earlier sections illustrate their ultimate impact on radiation exchange and energy balance. On the other hand, different soil management practices may be followed to manipulate hydraulic conductivity, thermal conductivity, as well as heat capacity of the soil. For example, a part of insolation received by the surface is absorbed and conducted downward as ground heat flux. Increasing the moisture content of soil (irrigation) increases the thermal conductivity and heat capacity of soil because water has higher thermal conductivity and higher heat capacity than air. Furthermore, the wet soil has a lower albedo than the dry soil. For all these reasons the application of irrigation enhances the heat storage of soil, and hence this practice is followed as a protection measure against cold injury or frost.

Another measure against cold injury is to provide an insulating cover (like polythene sheet, ash, or sawdust over the young seedlings in the nursery bed. Such an insulating layer restricts outgoing thermal radiation during long winter nights, thereby conserving the energy balance. Besides, the polythene cover or a shade tree restricts the formation of eddies over the crop surface and thus limits the convective loss of energy.

Way Forward

Microclimatic modifications help to develop a favorable environment in the vicinity of the plants, allowing optimum crop growth and development to ensure sustainable production. While climate change research mostly focuses on adaption and mitigation options, microclimates go largely unobserved and unattended. Under changing climate scenarios with an increased probability of extreme events, the judicious modification of microclimate may prove beneficial. Microclimate modification requires a complete knowledge of the physiology of plants and the physical environment. Various local practices for microclimate modification followed by traditional farmers of different regions need to be systematically documented, scientifically evaluated, and judiciously adopted to maintain production sustainability under changing climate scenarios.

References

Chen, J., Saunders, S.C., Crow, T.R., Naiman, R.J., Brosofske, K.D., Mroz, G.D., Brookshire, B.L., and Franklin, J.F. (1999). Microclimate in forest ecosystem and landscape ecology: Variations in local climate can be used to monitor and compare the effects of different management regimes. *BioScience*, 49(4), 288–297.

Gliessman, S.R. (2015). *Agroecology: The Ecology of Sustainable Food Systems* (3rd edn.). Taylor & Francis Group.

Kaur, J. and Singh, G. (2020). Adoption of microclimate modification techniques. *Just Agriculture*, 1(3), 1–7.

Managing the Microclimate. *Practical Note-27*. Spate Network Foundation. https://floodbased.org/documents/.

Stigter, C.J. (1994a). Management and manipulation of microclimate. In J.F. Griffiths (Ed.), *Handbook of Agricultural Meteorology*. Oxford University Press, pp. 273–284.

Stigter, C.J. (1994b). *Micrometeorology*. In J.F. Griffiths (ed.), *Handbook of Agricultural Meteorology*. Oxford University Press.

Wilken, G.C. (1972). Microclimate management by traditional farmers. *Geographical Review*, 62(4), 544–560.

Chapter 13

Modeling Crop Water Use for Enhanced Crop Productivity

Vyas Pandey

Department of Agricultural Meteorology,
Former Emeritus Scientist (ICAR),
Anand Agricultural University,
Anand, Gujarat, India
pandey04@yahoo.com

Abstract

The irrigation management program in agriculture demands estimation of crop water use at the field level, which aims to quantify the amount of water needed to replenish the depleted water in the crop root zone as a result of evaporative water loss called evapotranspiration (ET). The crop water needs to vary with the crop growth, mainly due to variations in crop height, green biomass, and environmental conditions. The cropping technique and irrigation methods also influence the crop water requirement. The estimation of crop evapotranspiration (ETc) on a daily basis for the crop growth period may be considered equal to the water requirement of the given crop because most of the water uptake by plants from soil is lost as evapotranspiration (ET). Crop evapotranspiration can be measured by Lysimeter, but such instruments are not feasible to install in desired fields. Hence, various modeling approaches are suggested for direct or indirect estimation of evapotranspiration. Among various methods, the Penman–Monteith method has been universally accepted and widely used to calculate reference evapotranspiration (ETo). The crop evapotranspiration or crop water requirement (ETc) is obtained by multiplying the reference evapotranspiration (ETo) with a crop coefficient (Kc) factor.

Keywords: Crop water use, Crop coefficient (Kc), Crop growth stages, Crop evapotranspiration, Crop water requirement (ETc), Penman-Monteith method, FAO-56, Reference evapotranspiration (ETo).

Introduction

Water is one of the natural resources that play an important role in the living world. Due to the uneven distribution of water around the globe, the proper management of water resources becomes extremely important. There is a great water demand for agriculture, industry, domestic, and all other sectors. The agricultural sector uses a major share of available water, hence, it is essential to improve water management in agriculture. The adoption of exact or correct amount of water and correct timing of application is essential in irrigation scheduling for optimum crop production. The accurate estimation of irrigation water requirements can save water in the agriculture sector.

The crop water needs vary with the crop growth, mainly due to variations in crop height, green biomass, and environmental conditions. The cropping technique and irrigation methods also influence the crop water requirement. The estimation of crop evapotranspiration (ETc) on a daily basis for the crop growth period may be considered equal to the water requirement of the given crop because most of the water uptake by plants from soil is lost as evapotranspiration (ET).

The evapotranspiration varies with variations in climatic parameters and also with the types of crop/vegetation (Allen *et al.*, 1998). Jensen *et al.* (1990) have given details of various methods to determine evapotranspiration and crop water requirements. The irrigation scheduling based on crop water requirement (ETc) determined by using crop coefficient (Kc) and reference evapotranspiration (ETo), is one of the widely used methods (Doorenbos and Pruitt, 1977). The crop water requirement depends on the age of the crop, crop growing season, its location, and management strategies to be adopted, and their computation needs information on reference crop evapotranspiration, crop coefficient, etc. Absence of this information may lead to either under or over-application of water. Among different methods for estimating evapotranspiration rates, the climatological-based methods are widely used.

This chapter describes the various methods used for estimating reference evapotranspiration (ETo), its spatial and temporal variation, determining the corrected Kc values, and thereby ETc for different crops including horticultural crops grown under different environments and also under micro-irrigation systems.

Estimation of Evapotranspiration

To estimate actual evapotranspiration, the term potential evapotranspiration (PET) is defined, which is the maximum evapotranspiration from a vegetation completely covering the ground surface that has unlimited water supply to its roots (Thornthwaite, 1948). Another term, reference evapotranspiration (ETo) is defined as the loss of water to the atmosphere by evaporation and transpiration from an extended surface of 8–12 cm tall green grass cover, usually well-watered, actively growing, and completely shading the ground. Although there is a slight difference between the two terms, PET and Eto, these terms have been used by many researchers for the same purpose (Khandelwal *et al.*, 1999, 2008). The empirical estimation of reference evapotranspiration (ETo) requires climatic data particularly on temperature, relative humidity, wind speed, and solar radiation.

Evapotranspiration can be measured by Lysimeter but such instruments are not feasible to install in desired fields. Hence, various approaches are suggested for direct or indirect estimation of evapotranspiration. The simplest one is as residual of water balance methods where precipitation/irrigation is taken as an input variable and soil moisture storage and run off/deep percolation as output are measured, giving the balance as ET. More advanced techniques such as micro-meteorological take into account the instantaneous variation/fluctuation in weather elements such as the Bowen ratio, energy balance method (Bowen, 1926), and eddy covariance method (Swinbank, 1951). The large aperture scintillometers have also been used to quantify the surface energy fluxes over a region. The latent heat flux was calculated as the residual of the energy balance (Nigam *et al.*, 2008).

Though such methods give a more accurate estimation of evaporation and/or transpiration, these cannot be applied in the field for determining crop water requirement and irrigation scheduling.

Moreover, these methods require more sophisticated instrumentations. Hence, climatological approaches are widely accepted and used. There are different methods for estimating potential evapotranspiration (PET)/reference evapotranspiration (ETo). Based on the requirement of climatic parameters, these are grouped into temperature, pan evaporation, radiation, and energy balance-based methods as described in the following sections.

Temperature-based methods

The simplest method to compute the potential evapotranspiration (ETo) was developed by Thornthwaite (1948). The equation is given as:

$$\text{ETo} = 1.6(10\ T/I)^a * (N/12) * (m/30), \tag{1}$$

where, "a" is a constant that varies with heat index I $= \sum_{i=1}^{12} (T/5)^{1.514}$.

Other parameters have their usual meaning as given in the Appendix. This method has been widely used across the globe.

Another mathematical relationship for computing ETo was developed by Blaney and Criddle (1950) using air temperature and daylight factors as:

$$\text{ETo} = k(p(0.46T + 8.13)). \tag{2}$$

The accuracy of ET estimation on a daily basis has limitation, hence suitable for longer time period ET estimation (Wright, 1985).

Hargreaves (1975) developed an equation to estimate potential evapotranspiration (ETo) using temperature, relative humidity, and latitude:

$$\text{ETo} = M_f(1.8T + 32)C_H. \tag{3}$$

Pan evaporation-based methods

Christiansen (1968) developed an empirical equation to estimate ETo using temperature, wind speed, humidity, solar radiation, and pan evaporation. The equation is of the form:

$$\text{ETo} = 0.755\,\text{E}_{\text{pan}} * C_T * C_W * C_H * C_S. \tag{4}$$

Another method for estimating daily ETo was developed by Doorenbos and Pruitt (1977) using pan coefficient. The equation is given as:

$$\text{ETo} = K_p * \text{E}_{\text{pan}}. \tag{5}$$

Radiation-based methods

Since solar radiation is the main source of energy for governing physical processes in the atmosphere and at the Earth's surface, several methods have been developed based on solar radiation and also involving other parameters such as temperature, wind, and other terms.

Makkink (1957) developed an equation to estimate ETo using incoming short-wave radiation R_s for grasslands in Holland. The equation is given as:

$$\text{ETo} = 0.65 \frac{\Delta}{\Delta + \gamma} (R_s - G). \tag{6}$$

Another equation using average daily radiation and temperature to estimate ETo given by Turc (1961) is expressed as:

$$\text{ETo} = 0.013(23.88^* R_s + 50)T(T + 15)^{-1}. \tag{7}$$

It is the simplest and the most accurate method of ETo estimation for humid regions (Jensen *et al.*, 1990).

Priestley and Taylor (1972) developed an equation using temperature and net radiation data to calculate ETo and is expressed as:

$$\text{ETo} = \frac{\alpha \Delta}{\lambda \Delta + \gamma} (R_n - G). \tag{8}$$

Doorenbos and Pruitt (1977) developed another equation to compute ETo using air temperature and radiation data.

$$\text{ETo} = c(0.408\text{W}^* R_s). \tag{9}$$

The constants "c" varies with relative humidity and wind speed and "W" varies with temperature and altitude. This is a useful method for ETo estimation (7) as it depends on temperature and sunlight hours (Chiew *et al.*, 1995).

Hargreaves and Samani (1985) developed another equation using temperature and extraterrestrial solar radiation data.

$$\text{ETo} = 0.0023(T + 17.8)R_a\sqrt{(\text{T}_{\max} - \text{T}_{\min})}. \tag{10}$$

Energy balance-based methods

Penman (1948) developed a method using the concept of energy supply and transport of water vapor from the surface. The evaporation from water surface (Eo) can be estimated using the equation;

$$\text{Eo} = \frac{\frac{\Delta * R_n}{\lambda} + \gamma * \text{Ea}}{\Delta + \gamma}. \tag{11}$$

Doorenbos and Pruitt (1977) modified the aforementioned equation to account it for a grass surface instead of a water surface. The equation is given as:

$$\text{ETo} = \text{C}\left[\frac{\Delta}{\Delta + \gamma} * R_n + \frac{\Upsilon}{\Delta + \gamma} * (0.27)(1 - 0.1u_2)(e_s - e_a)\right]. \tag{12}$$

Monteith (1965) introduced canopy resistance (r_c) term in Penman's method to estimate ETo. This method has been recommended by FAO (Doorenbos and Pruitt, 1977). The FAO P-M equation is given as:

$$\text{ETo} = \frac{0.408\Delta(R_n - G) + \gamma\frac{900}{\text{T}_a+273}u_2(e_s - e_a)}{\Delta + \gamma(1 + 0.34u_2)}. \tag{13}$$

The detailed step-wise procedure to compute daily reference evapotranspiration (Equation (13)) is described by FAO (Doorenbos and Pruitt, 1977) and Mehta and Pandey (2015, 2018).

Variability and Trends in Reference Evapotranspiration (ETo)

As described earlier, all the methods used for estimating the reference evapotranspiration (ETo) or potential evapotranspiration (PET) incorporated one or more climatic parameters, *viz.*, solar radiation, vapor pressure deficit, wind speed, and air temperature. The basic principles behind these methods are that these variables determine the energy required for evaporation to take place and also the transportation of water vapor. The FAI P-M method consists of two main components, namely energy balance component and aerodynamic term component. The contribution of each component in total ETo value may vary with climatic regions of the world. The energy balance component contributed 70–75% of ETo in India (Rao and Wani, 2011; Chakravarty *et al.*, 2015). Due to global warming and climate changes the atmospheric parameters are changing over a period of time, hence ETo is also bound to vary and change with location, season, and time.

Spatial variability

The annual ETo computed by FAO Penman–Monteith method for 18 stations of Gujarat presented in Fig. 1 shows that annual ETo in Gujarat varies from 1,912 mm to 3,041 mm. The highest ETo (3,041 mm) is at Arnej which is part of middle Gujarat, followed by 2,937 mm at Targadia and 2,847 mm at Amreli of Saurashtra region, and 2,805 mm at Viramgam of north Gujarat. The lowest value of annual ETo (1,912 mm) is observed at Khedbrahma, which is the northern part of Gujarat, followed by 2,101 mm at Paria, 2,294 mm at Navsari, and 2,343 mm at Surat all from south Gujarat and 2,242 mm

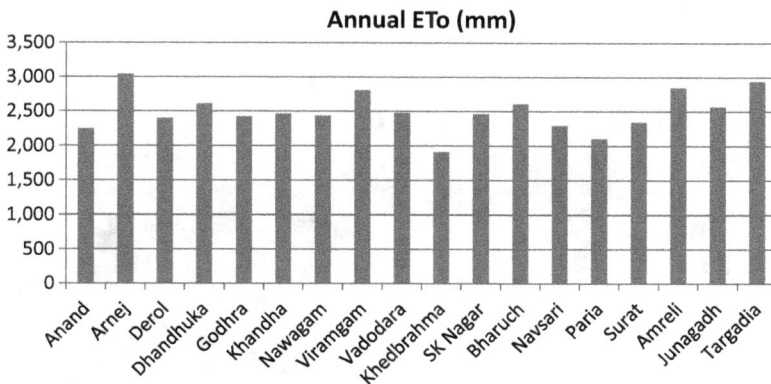

Fig. 1. Variation of annual ETo in Gujarat.

at Anand in middle Gujarat. The variation in ETo is mainly due to variations in the climatic conditions of the stations.

Temporal variability

The monthly ETo of selected stations of Gujarat presented in Fig. 2 shows that during a year ETo varies during different months. ETo is highest during the summer season and lowest during the winter season. From January to May the ETo increases continuously, the rate of increase being different in different months with the peak ETo being in May at all the stations. The ETo decreases sharply during June and July due to an increase in humidity as a result of the onset of monsoon then a slight decrease during August. It further increases slightly during September and October before reaching its lowest value during December. These variations are not uniform at all locations because the energy balance and aerodynamic components of ETo estimation vary with location. Nag *et al.* (2014) have also found that May had maximum ETo and December and January had minimum ETo for most of the states and basins of India.

Trends

There are reports that evapotranspiration rates have been changing over the past several years. In various parts of the world, a decreasing

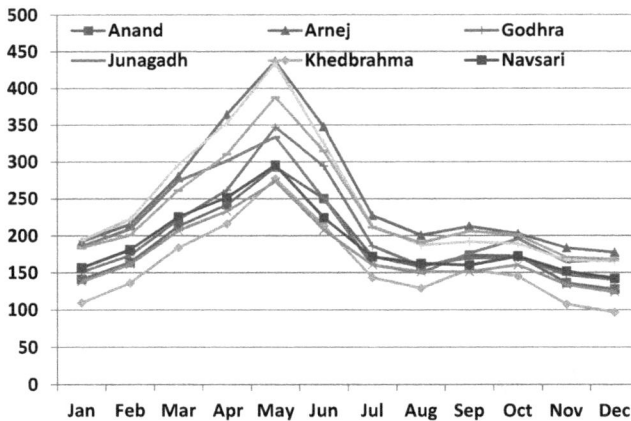

Fig. 2. Monthly variation of ETo (mm) at selected stations of Gujarat.

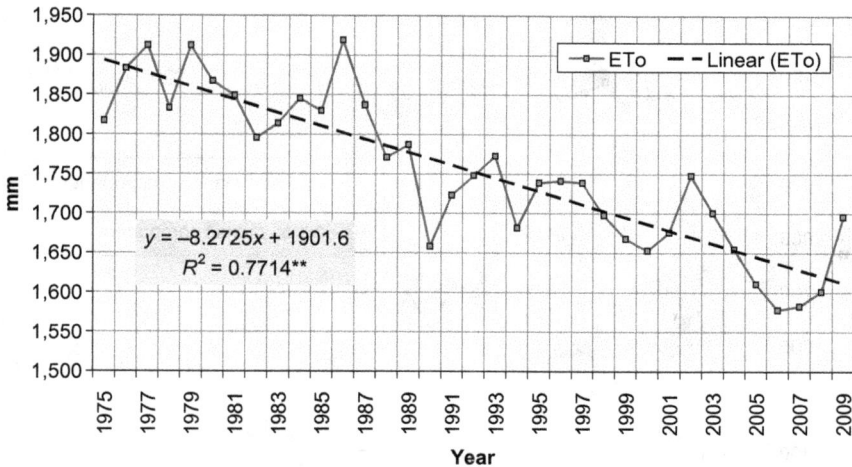

Fig. 3. Trend in reference crop evapotranspiration (ETo) at ICRISAT, Patancheru.

Source: Rao and Wani (2011).

trend in evapotranspiration is observed. Under the projected climate change scenarios with increased temperatures, the implications of the evapotranspiration trends on the hydrological cycle are somewhat controversial. The paradox of decreasing evaporation and evapotranspiration demands under increasing temperature conditions exists. Figure 3 shows a significant decreasing trend in annual reference crop evapotranspiration at ICRISAT, Patancheru, Hyderabad during a 35-year (1975–2009) period (Rao and Wani, 2011). Average annual ETo decreased by 57 mm per decade or about 3% of the annual total. Changes in the two components (energy balance and aerodynamic) of the ETo were examined to better understand the conspicuous decrease in the ETo and it is seen that the energy balance component showed a positive trend while the aerodynamic component showed a highly significant negative trend (Fig. 4). Annual energy component has increased from about 1,000 mm to 1,100 mm while aerodynamic component decreased from about 800 mm to 550 mm. Rate of increase for the energy balance component is about 29 mm per decade while the rate of decrease for the aerodynamic component is about 71 mm per decade.

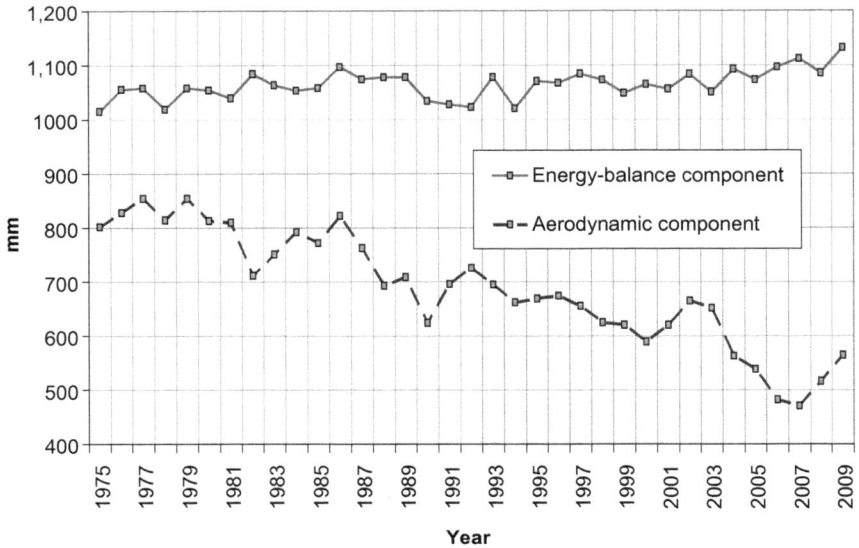

Fig. 4. Trend in energy balance and aerodynamic components of ETo at ICRISAT, Patancheru.
Source: Rao and Wani (2011).

Crop Coefficient and Irrigation Requirement

The crop coefficients are determined experimentally while measuring crop water requirement using Lysimeter (Rowshon *et al.*, 2013). To determine the crop water requirements (ETc) the crop coefficient (Kc) and reference evapotranspiration (ETo) are used (Doorenbos and Pruitt, 1977).

Crop coefficient

The crop coefficient is defined as the ratio between actual crop evapotranspiration (ETc) of the crop and the reference evapotranspiration (ETo). Since Lysimeters cannot be installed in each field area, it is convenient to apply Kc values of crop to obtain crop water requirement (ETc) from reference evapotranspiration (ETo). The value of Kc changes as the crop grows and crop canopy develops. A schematic diagram of the variation of Kc values with crop growth stages is given in Fig. 5. During the initial period of the crop growth stage if the

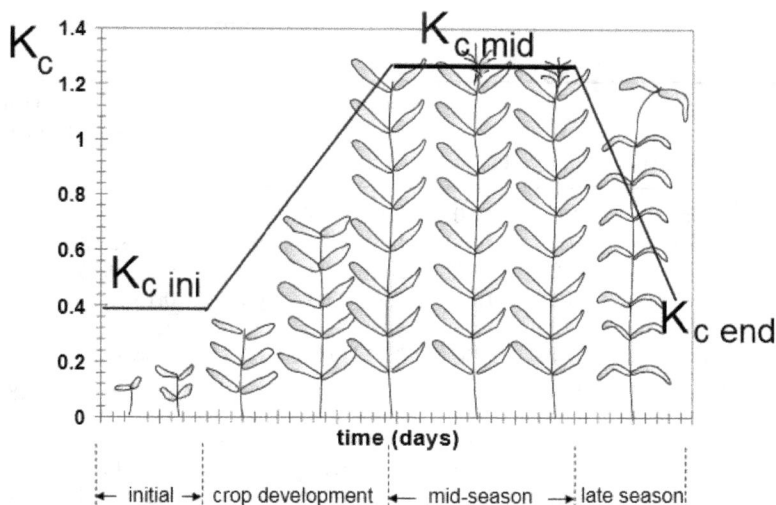

Fig. 5. Schematic diagram of variation of Kc values with crop growth stages.

soil surface is not covered by the crop fully (groundcover <10%) then the Kc values are less. During the crop development stage till attaining full groundcover, the Kc value increases continuously. It remains more or less constant during the subsequent period till the start of maturity. Thereafter, Kc values decrease during the maturity period till the harvest of the crop.

The crop coefficient (Kc) value also varies with variations in climatic conditions during the crop-growing season. The crop coefficients are adjusted based on wind and humidity (Doorenbos and Pruitt, 1977). The crop coefficient (Kc_{mid} and Kc_{end}) values for different crops were corrected according to climatic conditions of different locations in Gujarat, whereas, the Kc values for the developmental stage and late season stage for each of the crops were calculated by linear interpolation (Mehta and Pandey, 2018).

$$\text{Kc}_{\text{mid}} = \text{Kc}_{\text{mid(tab)}} + [0.04(\text{u}_2 - 2) - 0.004(\text{RH}_{\text{min}} - 45)]\left[\frac{\text{h}}{3}\right]^{0.3},$$

$$\text{Kc}_{\text{end}} = \text{Kc}_{\text{end(tab)}} + [0.04(\text{u}_2 - 2) - 0.004(\text{RH}_{\text{min}} - 45)]\left[\frac{\text{h}}{3}\right]^{0.3}.$$

The symbols have their usual meanings.

Table 1. Corrected crop coefficients (Kc) of different crops along with that given as per FAO.

S. No.	Crops	Kc_{ini}	Kc_{mid}	Corrected Kc_{mid}	Kc_{end}	Corrected Kc_{end}	Plant height (m)
1.	Rice	1.05	1.20	1.11–1.20	0.90	0.90	0.8
2.	Wheat	0.40	1.15	1.13–1.17	0.41	0.39–0.43	0.8
3.	Maize	0.30	1.20	1.08–1.20	0.35	0.23–0.35	1.5
4.	Pearl millet	0.30	1.00	0.87–1.00	0.30	0.18–0.24	1.5
5.	Chickpea	0.40	1.00	0.67–1.01	0.35	0.34–0.36	0.4
6.	Green gram	0.40	1.05	0.96–1.05	0.60	0.51–0.60	0.4
7.	Soybean	0.40	1.15	1.05–1.15	0.50	0.41–0.50	0.6
8.	Groundnut	0.40	1.15	1.06–1.15	0.60	0.51–0.60	0.4
9.	Mustard	0.35	1.15	1.13–1.16	0.35	0.33–0.36	1.3
10.	Cotton	0.35	1.15	1.08–1.15	0.50	0.43–0.50	1.1

Source: Allen *et al.* (1998) and Pandey and Mehta (2018).

The corrected crop coefficient (Kc) values determined for 10 major crops (rice, wheat, maize, pearl millet, chickpea, green gram, soybean, groundnut, mustard, and cotton) at different selected locations in Gujarat have been reported (Pandey and Mehta, 2018), and their values over the locations are presented in Table 1. Kc_{mid} and Kc_{end} of most of the crops were found to vary across the locations. Generally in dry regions corrected Kc values are higher than the FAO Kc values. Moreover, most of the locations in north Saurashtra and Bhal regions recorded higher Kc values.

Crop coefficient of horticultural crops: Williams and Ayars (2005) developed a linear relationship between crop coefficients of grape wines grown under Lysimeter with leaf area, leaf area index (LAI), and percent shaded area. Among all the three relations developed, one with percent shaded area explained the crop coefficient maximum. Such relationship was used to determine the crop coefficient and water requirement for **pomegranate** (*Punica granatum* L.) under different management practices in Maharashtra, India (Gorantiwar *et al.*, 2011; Meshram *et al.*, 2018).

$$Y = -0.008 + 0.017X, \quad R^2 = 0.95^{**},$$

where Y is crop coefficient (Kc) and X is the percent shaded area under the deciduous fruit crops.

Gadge *et al.* (2011) developed equations to determine the daily crop coefficients of several crops including horticultural crops of Maharashtra under micro irrigation system as a function of day and duration of the crops.

Crop water requirement

As described earlier, there are different methods to estimate the ETc using climatic factors. To determine actual crop evapotranspiration (ETc) two factors are used, *viz.*, reference crop evapotranspiration and crop coefficient.

$$ETc = (Kc)(ETo),$$

where:

ETc = actual crop evapotranspiration rate
Kc = crop coefficient
ETo = evapotranspiration rate for a grass reference crop.

Water requirement under micro irrigation system: Gorantiwar *et al.* (2011) and Meshram *et al.* (2018) determined the water requirement of pomegranate under the drip irrigation method which is less than the water requirement under surface irrigation system. The water requirement (W_R) under drip irrigation systems are estimated by:

$$W_R = Fa^*ETc,$$

where Fa = Area factor (fraction) which is calculated as the ratio of shaded area divided by the area occupied by the tree.

Water Requirement (ETc) of Different Crops

The crop water requirements are estimated by combining crop coefficient (Kc) and reference evapotranspiration (ETo), both of which vary with the climatic conditions of the area as described in previous sections. Moreover, the crop coefficient varies with the growth stage, height, and duration of the crop. The crop water requirement also varies with the stage of the crop, its location, and climatic conditions. The crop water requirement that worked out in India for some crops are presented here.

(a) *Rice*: Rice is grown during the *kharif* season in the northern part of India. It is also grown during winter and summer seasons in the southern and eastern states. ETc of *kharif* rice at Sabour and Patna in Bihar was 546 and 607 mm, respectively (Kumar, 2017). Khandelwal and Dhiman (2015) reported ETc of 640 mm for *kharif* rice and 851 mm for summer rice in Mahi canal command area of Gujarat. Figure 6 shows the crop water requirement (ETc) of *kharif* rice during different growth stages over different stations in Gujarat. ETc varied between 70 and 95 mm during the initial stage, 139 and 195 mm during the developmental stage, 266 and 315 mm during mid-season stage, and 135 and 165 mm during the late season with a seasonal variation of 618 and 754 mm in Gujarat.

(b) *Wheat*: Wheat is grown during the winter (*Rabi*) season in most parts of India. Figure 7 shows the crop water requirement (ETc) of wheat during different growth stages across Gujarat. ETc varied between 27 and 48 mm during the initial stage, 73 and 135 mm during the developmental stage, 201 and 351 mm during mid-season stage, and 97 and 156 mm during the late season with seasonal totals of 398–680 mm across locations in

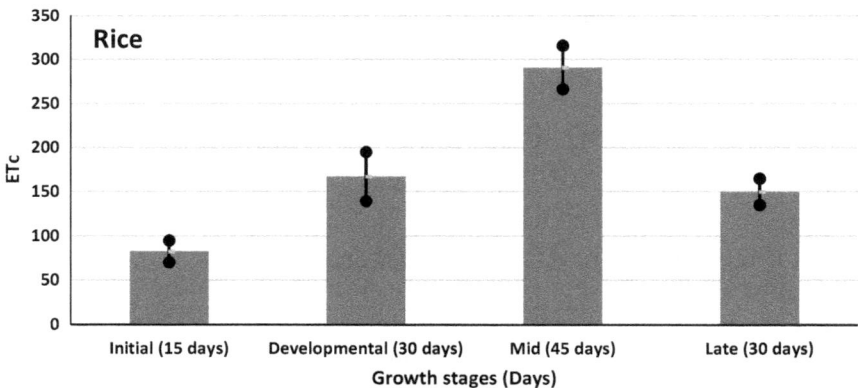

Fig. 6. Crop water requirement during different growth stages of *kharif* rice. Bar shows its variation across the locations in Gujarat.

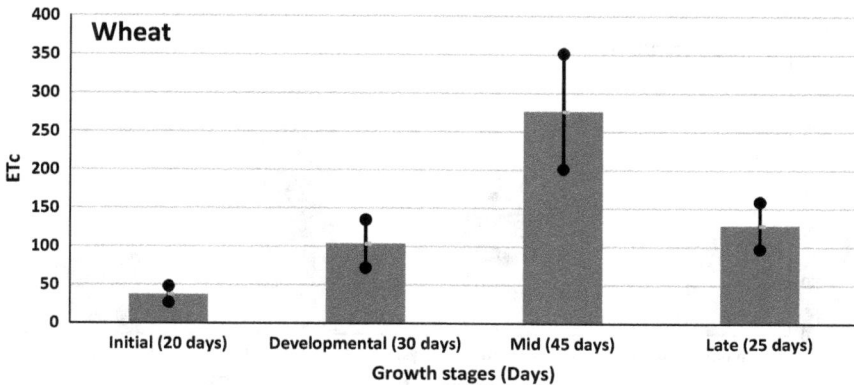

Fig. 7. Crop water requirement during different growth stages of wheat. Bar shows its variation across the locations in Gujarat.

Gujarat. Kumar (2017) reported seasonal ETc of wheat as 213 and 243 mm at Sabour and Patna, respectively, while Khandelwal and Dhiman (2015) reported 565 mm in Mahi canal command area of Gujarat. Rao and Poonia (2011) reported ETc of wheat in Rajasthan ranging between 173 and 288 mm.

(c) *Maize*: Maize is grown in all seasons in different parts of India. Kumar (2017) determined ETc of maize in all three seasons (*kharif, rabi*, and summer) in Bihar and reported that it was between 292 and 319 mm during *kharif* season, between 323 and 372 mm during the *rabi* season, and between 500 and 591 mm during the summer season. In contrast to this, it is reported that the ETc of maize during the *rabi* season was slightly less than that of the *kharif* season at Anand Gujarat (Mehta and Pandey, 2016). Figure 8 shows the crop water requirement (ETc) of *rabi* maize during different growth stages over different stations in Gujarat. ETc was found to vary between 15 and 22 mm during the initial stage, 69 and 110 mm during the developmental stage, 174 and 282 mm during mid-season stage, and 72 and 107 mm during the late season with seasonal variation of 331–521 mm in Gujarat.

(d) *Pearl millet*: The crop water requirement (ETc) of *kharif* pearl millet during its different growth stages over different stations of Gujarat are presented in Fig. 9. ETc was found to vary between 20 and 31 mm during the initial stage, 66 and 93 mm during the

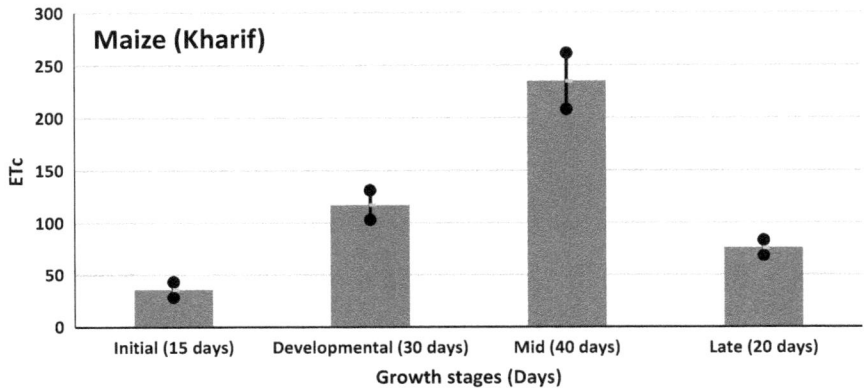

Fig. 8. Crop water requirement during different growth stages of maize. Bar shows its variation across Gujarat.

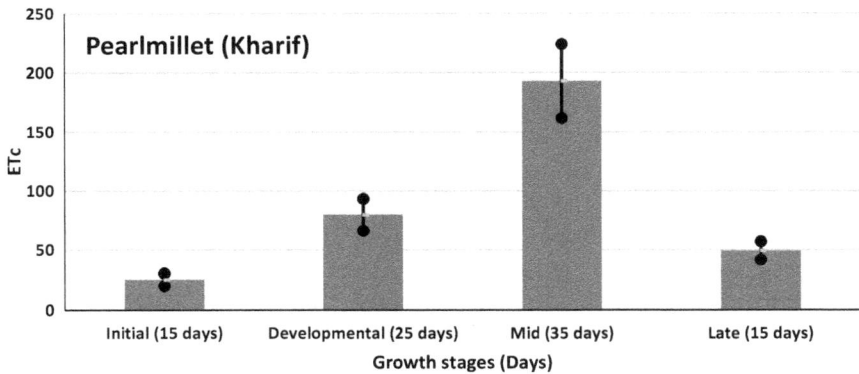

Fig. 9. Crop water requirement during different growth stages of pearl millet. Bar shows its variation across Gujarat.

developmental stage, 162 and 224 mm during mid-season stage, and 42 and 57 mm during the late season with seasonal variation of 305–402 mm in Gujarat. Similar range of ETc of *kharif* pearl millet as 25, 88, 124, and 47 mm during initial stage, developmental stage, mid-season stage, and late season, respectively, in central Gujarat was reported by Rao *et al.* (2012). ETc of summer pearl millet (499 mm) was more than that of *kharif* pearl millet (324 mm) at Anand (Mehta and Pandey, 2016). Even higher ETc of 619 mm for summer pearl millet in Mahi canal command area of Gujarat (18) has been reported while Rao and Poonia (2011) reported a wide variation in ETc of *kharif* pearl millet in western Rajasthan ranging between 308 and 411 mm.

(e) *Soybean*: Figure 10 shows the crop water requirement (ETc) of soybean during its different growth stages with its variation across the locations in Gujarat. ETc was found to vary between 49 and 76 mm during the initial stage, 103 and 148 mm during the developmental stage, 236 and 315 mm during mid-season stage, and 97 and 131 mm during the late season with seasonal variation of 490–670 mm in Gujarat. ETc of soybean at Anand was 533 mm (Mehta and Pandey, 2016). In Bangladesh, ETc of soybean was 35, 131, 162, and 51 mm during initial, developmental, mid-season, and late season, respectively (Mila *et al.*, 2016), which are less than that reported in the present study (Fig. 10).

(f) *Groundnut*: Groundnut is grown both during the *kharif* and summer seasons. The crop water requirement (ETc) of *kharif* groundnut during its different growth stages over different stations of Gujarat are presented in Fig. 11. ETc of *kharif* groundnut was found to vary between 49 and 76 mm during the initial stage, 117 and 172 mm during the developmental stage, 213 and 281 mm during mid-season stage and 105 and 140 mm during the late season with seasonal variation of 488–670 mm in Gujarat. ETc of summer groundnut (849 mm) was more than that of *kharif*

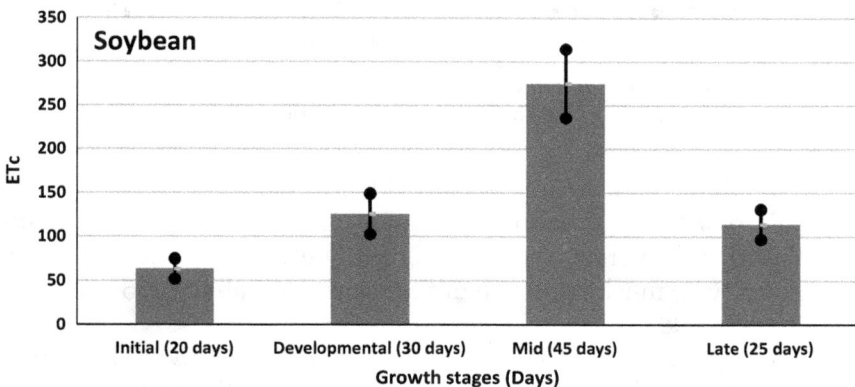

Fig. 10. Crop water requirement during different growth stages of soybean. Bar shows its variation across the locations in Gujarat.

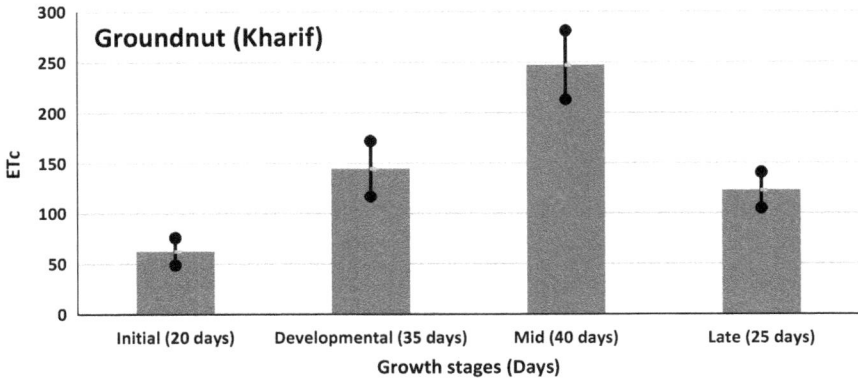

Fig. 11. Crop water requirement during different growth stages of groundnut. Bar shows its variation across the locations in Gujarat.

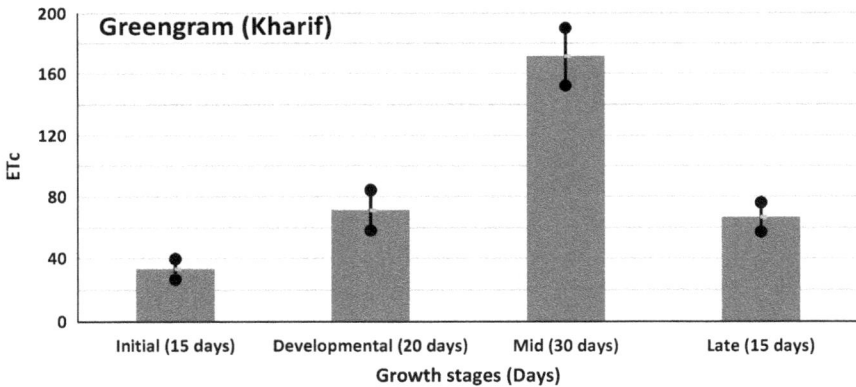

Fig. 12. Crop water requirement during different growth stages of green gram. Bar shows its variation across the locations in Gujarat.

groundnut (536 mm) at Anand (Mehta and Pandey, 2016). The ETc of *kharif* groundnut at Anantpur, Andhra Pradesh, estimated by different methods ranged between 281 and 434 mm which were much higher than the measured values (455–600 mm) (Reddy, 1988).

(g) *Green gram*: Green gram is also grown during *kharif* and summer seasons. The crop water requirement (ETc) of *kharif* green gram during its different growth stages over different stations of Gujarat (Fig, 12) is less than that of groundnut (Fig. 11). ETc of *kharif* green gram was found to vary between 27 and

40 mm during the initial stage, 58 and 84 mm during the developmental stage, 152 and 190 mm during mid-season stage, and 57 and 76 mm during the late season with seasonal variation of 299–387 mm in Gujarat. Large variations in ETc of *kharif* green gram in western Rajasthan ranging between 216 and 297 mm have been reported (Rao and Poonia, 2011). Kumar (2017) reported that ETc of summer green gram was 405 mm at Sabour and 476 mm at Patna in Bihar.

(h) *Chickpea*: Chickpea is grown during the winters (*Rabi*) in most parts of India. Figure 13 shows the crop water requirement (ETc) of chickpea during different growth stages over different stations in Gujarat. ETc of chickpea varied between 23 and 37 mm during the initial stage, 79 and 144 mm during the developmental stage, 155 and 276 mm during mid-season stage, and 77 and 125 mm during the late season with seasonal totals of 334–581 mm across the locations in Gujarat. In Ethiopia, the ETc of chickpea during initial, developmental, mid-season, and late season were 37, 114, 205, and 80 mm, respectively (Desta *et al.*, 2015), which are similar to the reported values in the present study (Fig. 13).

(i) *Cluster bean*: Cluster bean is grown in dry regions of India. Rao and Poonia (2011) worked out the crop water requirement of

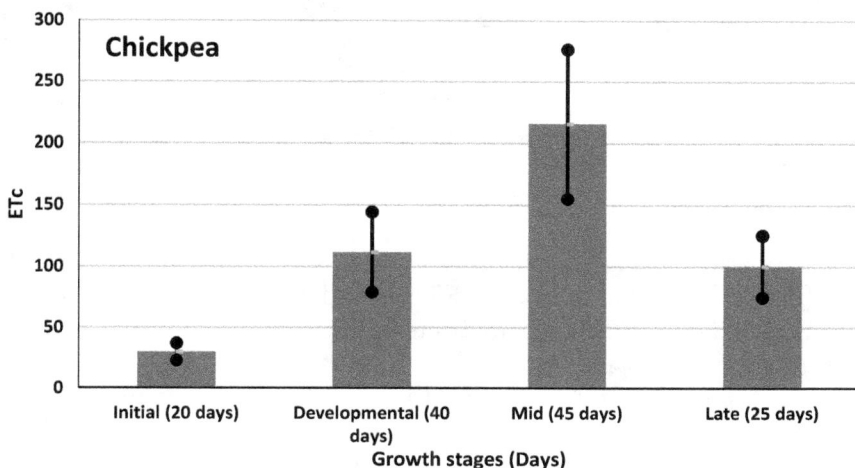

Fig. 13. Crop water requirement during different growth stages of chickpea. Bar shows its variation across the locations in Gujarat.

Cluster bean ETc (mm)

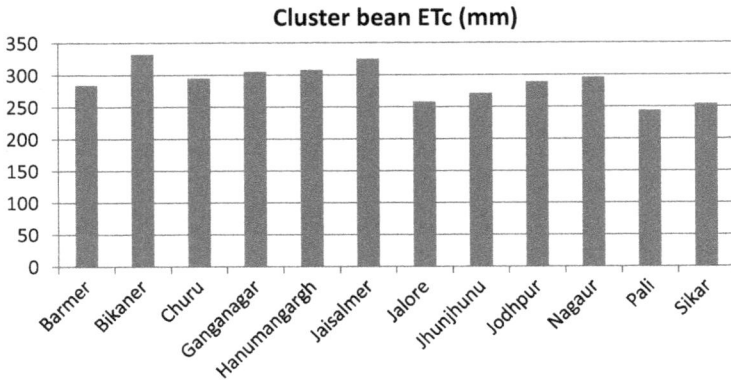

Fig. 14. Seasonal crop water requirement of cluster beans in western Rajasthan. *Source*: Rao and Poonia (2011).

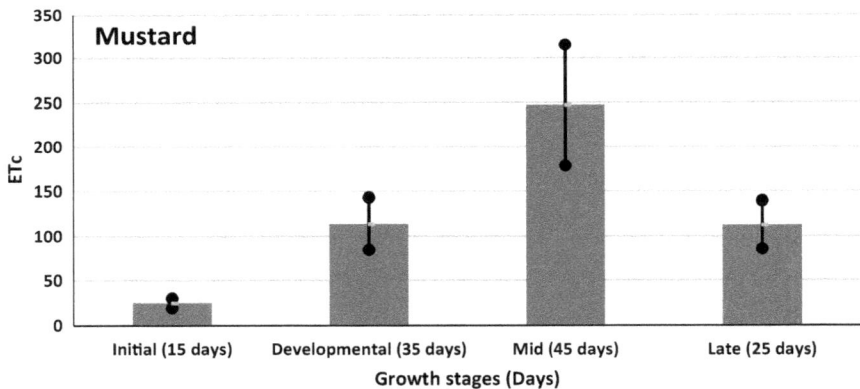

Fig. 15. Crop water requirement during different growth stages of mustard. Bar shows its variation across the locations in Gujarat.

cluster beans in different districts of western Rajasthan (Fig. 14). The figure shows that the ETc varied between 244 mm in Pali and 332 mm in Bikaner districts of Rajasthan.

(j) *Mustard*: Mustard is grown during the winter (*rabi*) season in most parts of India. Figure 15 shows the crop water requirement (ETc) of mustard during different growth stages and its variation across the locations in Gujarat. ETc of mustard varied between 20 and 30 mm during the initial stage, 85 and 143 mm during the developmental stage, 178 and 315 mm during mid-season stage, and 85 and 139 mm during the late season with seasonal totals of 368–625 mm across the locations in Gujarat. In western Rajasthan, a large variation in ETc of mustard ranging between 214 and 343 mm has been reported (Rao and Poonia, 2011), but these values were less than that reported for Gujarat (Fig. 15). Khavse *et al.* (2014) also reported similar results in Chhattisgarh where seasonal ETc of mustard varied between 328 and 373 mm.

(k) *Cotton*: Cotton is a medium- to long-duration crop grown during the extended *kharif* season in most of India. Figure 16 shows the crop water requirement (ETc) of cotton during different growth stages and its variation across Gujarat. ETc of cotton varied between 59 and 90 mm during the initial stage, 137–197 mm during the developmental stage, 365 and 481 mm during mid-season stage, and 141–219 mm during the late season with seasonal totals of 702 and 986 mm across the locations in Gujarat. FAO has also reported that the water needs for cotton ranging between 700 and 1,300 mm.

(l) *Tobacco:* Tobacco is a medium-duration crop of 120–150 days. The water requirements (ETc) for maximum yield vary from 400 to 600 mm based on the climate and length of the growing period. The duration of different stages of tobacco is 30 days for the initial stage, 40 days for the developmental stage, 50 days

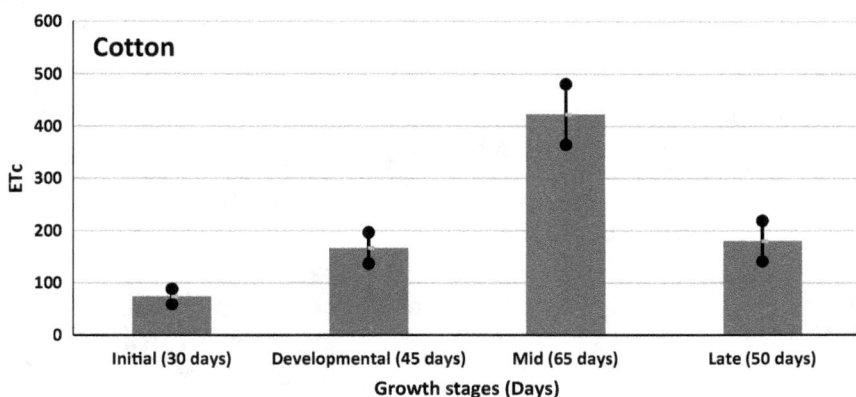

Fig. 16. Crop water requirement during different growth stages of cotton. Bar shows its variation across the locations in Gujarat.

for the mid-season stage, and 30 days for the late season. Rao
et al. (2012) worked out the ETc of tobacco in central Gujarat
and reported ETc of 47, 91,133, and 43 mm during the initial
stage, developmental stage, mid-season stage, and late season
respectively with a seasonal total of 314 mm only.

(m) *Saffron*: Saffron (*Crocus sativus* L.) is a perennial, herbaceous
plant that has been cultivated for its spice. The dried stigma
of this plant composes the most expensive spice in the world.
It is an important cash crop of Jammu and Kashmir state of
India. Ahmad *et al.* (2017b) worked out the water requirement
of saffron and reported that the total water requirement for the
saffron crop was 288 mm which is split into 80 mm during the
initial stage (sprouting to flowering), 134 mm during the mid-
season stage (vegetative growth period), and 74 mm during the
late season stage.

(n) *Apple*: Apple is a rosaceous fruit tree, majorly grown in states like
Jammu and Kashmir, Himachal Pradesh, Uttarakhand, Assam,
and Nilgiri Hills. Ahmad *et al.* (2017a) worked out the water
requirement of apples in Kashmir Valley and reported that the
mean ETc were 69, 668, and 175 mm during the initial, mid-
season, and late-season stages, respectively. Figure 17 shows the
seasonal ETc across the locations in Kashmir Valley.

(o) *Pomegranate*: Pomegranate is largely cultivated in marginal
lands with fertigation system and bahar treatment for regulating flowering and fruiting. Pomegranate is sensitive fruit crop
to water stress. The water requirement of pomegranate crops

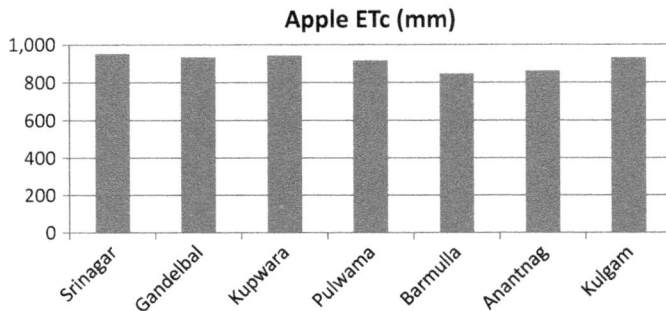

Fig. 17. Seasonal crop water requirement of apples in Kashmir Valley.
Source: Ahmad *et al.* (2017a).

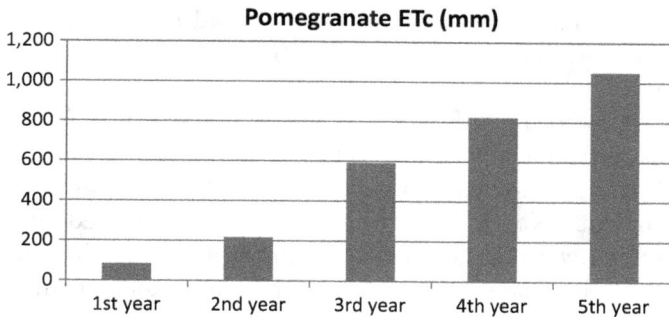

Fig. 18. Annual crop water requirement of pomegranate at different ages of the tree.
Source: Gorantiwar *et al.* (2011).

depends on age, season, location, and management strategies. Gorantiwar *et al.* (2011) and Meshram *et al.* (2018) worked out the ETc of pomegranates of different ages and reported that it increased with the age of the tree (Fig. 18).

Gadge *et al.* (2011) worked out the crop water requirement of several vegetable and horticultural crops of Maharashtra under micro irrigation system based on 95% efficiency and reported that the maximum water saving was in pomegranate and lime (88%) followed by papaya (75%) and grapes (73%) while the minimum saving was in summer groundnut (38%) followed by summer onion (40%) and *kharif* groundnut (43%).

Summary

It can be summarized that there are various approaches to estimate and/or measure the crop water requirement. Widely used and accepted practical approach is to estimate reference evapotranspiration (ETo) and apply crop coefficient (Kc) to obtain daily crop water requirement (ETc). There are several methods for estimating ETo, each one having its own merits and demerits. Penman–Monteith method was found universally acceptable as it is based on a complete scientific footing. FAO publication (Allen *et al.*, 1998) on crop water requirement gives a complete description of the methodology for estimating ETo, crop coefficient, and thereby crop water requirement.

The crop coefficients of different crops are also given in this publication, however, these values need to be corrected for local climatic conditions.

ETo estimated for different stations of Gujarat was found to vary temporally and spatially as well. The energy balance term contributes more to ETo variation in comparison to the aerodynamic term. The crop coefficient (Kc) for different crops was corrected at different stations which were found to differ slightly with the Kc values given in the FAO publication. The crop water requirement during different stages estimated for 11 crops of Gujarat were found to vary with stage as well as location. These results were supported by the works done elsewhere in the country. The crop water requirement of other crops including horticultural carried out elsewhere are also presented. The water requirement under micro-irrigation system was found to be less than the surface irrigation system.

References

Ahmad, L., Sabah, P., Saqib, P., and Kanth, R.H. (2017a). Reference evapotranspiration and crop water requirement of apple (*Malus Pumila*) in Kashmir Valley. *Journal of Agrometeorology*, 19(3), 262–264.

Ahmad, L., Sabah, P., Saqib, P., and Kanth, R.H. (2017b). Crop water requirement of saffron (*Crocus sativus*) in Kashmir Kanth Valley. *Journal of Agrometeorology*, 19(4), 380–381.

Allen, R.G., Pereira, L.S., Raes, D., and Smith, M. (1998). Crop evapotranspiration: Guidelines for computing crop water requirements. *Irrigation and Drainage Paper 56*. Food and Agriculture Organization of the United Nations, Rome, p. 300.

Blaney, H.F. and Criddle, W.D. (1950). Determining water requirements in irrigated areas from climatological and irrigation data. *USDA Soil Conservation Service*, 96, 44.

Bowen, I.S. (1926). The ratio of heat losses by conduction and by evaporation from any water surface. *Physical Review*, 27, 779–787.

Chakravarty, R., Bhan, M., Rao, A.V.R.K., and Awasthi, M.K. (2015). Trends and variability in evapotranspiration at Jabalpur, Madhya Pradesh. *Journal of Agrometeorology*, 17 (2), 199–203.

Chiew, F.H.S., Kamaladasa, N.N., Malano, H.M., and McMahon, T.A. (1995). Penman-Monteith, FAO-24 reference crop evapotranspiration and class-A pan data in Australia. *Agricultural Water Management*, 28, 9–21.

Christiansen, J.E. (1968). Pan evaporation and evapotranspiration from climatic data. *Journal of Irrigation and Drainage*, 94, 243–265.

Desta, F., Bissa, M., and Korbu, L. (2015). Crop water requirement determination of chickpea in central vertisol areas of Ethiopia using FAO CROPWAT model. *African Journal of Agricultural Research*, 10(7), 685–689.

Doorenbos, J. and Pruitt, W.O. (1977). Crop water requirements. *Irrigation and Drainage Paper 24*. Food and Agriculture Organization of the United Nations, Rome, p. 156.

Gadge, S.B., Gorantiwar, S.D., Kumar, V., and Kothar, M. (2011). Estimation of crop water requirement based on Penman-Monteith approach under micro-irrigation system. *Journal of Agrometeorology*, 13(1), 58–61.

Gorantiwar, S.D., Meshram, D.T., and Mittal, H.K. (2011). Water requirement of pomegranate (*Punica granatum* L.) for Ahmednagar district of Maharashtra State, India. *Journal of Agrometeorology*, 13(2), 123–127.

Hargreaves, G.H. (1975). Moisture availability and crop production. *Transactions of the ASAE*, 18(5), 980–984.

Hargreaves G.H. and Samani, Z.A. (1985). Reference crop evapotranspiration from temperature. *Applied Engineering in Agriculture*, 1, 96–99.

Jensen, M.E., Burman, R.D., and Allen, R.G. (1990). Evapotranspiration and irrigation water requirements. *ASCE Manuals and Reports on Engineering Practices*, 70, 360.

Khandelwal, M.K. and Pandey, V. (2008). Estimation of potential evapotranspiration computated by various methods in different agroclimatic stations of Gujarat state. *Journal Agrometeorology*, 10(Special issue-2), 439–443.

Khandelwal, M. K., Shekh, A. M., and Pandey, V. (1999). Selection of appropriate methods for computation of potential evapotranspiration and assessment of rainwater harvesting potential for middle Gujarat. *Journal Agrometeorology*, 1(2), 163–166. https://doi.org/10.54386/jam.v1i2.346.

Khandelwal, S.S. and Dhiman, S.D. (2015). Irrigation water requirements of different crops in Limbasi branch canal command area of Gujarat. *Journal of Agrometeorology*, 17(1), 114–117.

Khavse, R., Singh, R., Manikandan, N., Chandrawanshi, S.K., and Chaudhary, J.L. (2014). Crop water requirement and irrigation water requirement of mustard crop at selected locations of Chhattisgarh State, India. *Ecology. Environment & Conservation*, 20(Suppl.), S209–S211.

Kumar, S. (2017). Reference evapotranspiration (ETo) and irrigation water requirement of different crops in Bihar. *Journal of Agrometeorology*, 19(3), 238–241.

Makkink, G.F. (1957). Testing the Penman formula by means of lysimeters. *Journal of the Institution of Water Engineers*, 11(3), 277–288.

Mehta, R. and Pandey, V. (2015). Reference evapotranspiration (ETo) and crop water requirement (ETc) of wheat and maize in Gujarat. *Journal of Agrometeorology*, 17(1), 107–113.

Mehta, R. and Pandey, V. (2016). Crop water requirement (ETc) of different crops of middle Gujarat. *Journal of Agrometeorology*, 18(1), 83–87.

Mehta, R. and Pandey, V. (2018). *Crop Water Requirement — Evapotranspiration and Estimation*. Write and Print Publication, p. 106.

Meshram D.T., Gorantiwar, S.D., Sharma, J., and Babu, K.D. (2018). Influence of organic mulches and irrigation levels on growth, yield and water use efficiency of pomegranate (*Punica granatum* L.). *Journal of Agrometeorology*, 20(3), 196–201.

Mila, A.J., Akanda, A.R., and Sarkar, K.K. (2016). Determination of Crop Co-efficient Values of Soybean (*Glycine max* (L.) Merrill) by Lysimeter Study. *The Agriculturists*, 14(2), 14–23.

Monteith, J.L. (1965). Evaporation and environment. *Symposia Society Experimental Biology*, 19, 205–224.

Nag, A., Adamala, S., Raghuwanshi, N.S., Singh, R., and Bandyopadhyay, A. (2014). Estimation and ranking of reference evapotranspiration for different spatial scales in India. *Journal of Indian Water Resources Society*, 34(3), 35–45.

Nigam R., Mallick, K., Bhattacharya, B.K., Pandey, V., and Patel, N.K. (2008). Heat flux estimation from MODIS TERRA-AQUA and validation over a semi-arid agroecosystem using scintillometry and model simulation. *Journal of Agrometeorology*, 10(Special issue), 75–81.

Pandey, V. and Mehta, R. (2018). Reference evapotranspiration and water requirement of crops in Gujarat. Anand Agricultural University Report No. AAU/ICAR/Agmet/ES/Report-1, p. 94.

Penman, H.L. (1948). Natural evaporation from open water, bare soil and grass. *Proceedings of the Royal Society of London*, 193A, 120–146.

Priestley, C.H.B. and Taylor, R.J. (1972). On the assessment of surface heat flux and evaporation using large scale parameters. *Monthly Weather Review*, 100, 81–92.

Rao, A.S. and Poonia, S. (2011). Climate change impact on crop water requirements in arid Rajasthan. *Journal of Agrometeorology*, 13(1), 17–24.

Rao, A.V.R.K. and Wani, S.P. (2011). Evapotranspiration paradox at a semi-arid location in India. *Journal of Agrometeorology*, 13(1), 3–8.

Rao, B.K., Kumar, G., Kurothe, R.S., Pandey, V., Mishra, P.K., Vish-wakarma, A.K., and Baraiya, M.J. (2012). Determination of crop coefficients and optimum irrigation schedules for bidi tobacco and pearl millet crops in central Gujarat. *Journal of Agrometeorology*, 14(2), 123–129.

Reddy, P.R. (1988). Computing water requirement of groundnut. *Annals of Arid Zone*, 27(3 & 4), 247–251.

Rowshon, M.K., Amin, M.S.M., Mojid, M., and Yaji, M. (2013). Estimated evapotranspiration of rice based on pan evaporation as a surrogate to lysimeter measurement. *Paddy Water Environment*, 13(4), 356–364.

Singh, R., Singh, K., and Bhandarkar, D.M. (2014). Estimation of water requirement for soybean (*Glycine max*) and wheat (*Triticum aestivum*) under vertisols of Madhya Pradesh. *Indian Journal Agricultural Sciences*, 84(2), 190–197.

Swinbank, W. C. (1951). The measurement of vertical transfer of heat and water vapour by eddies in the lower atmosphere. Common wealth scientific research organization, Australia. *Journal Meteorology*, 8(3), 135–145.

Thornthwaite, C.W. (1948). An approach towards a rational classification of climate. *Geographical Review*, 38, 55–94.

Turc, L. (1961). Evaluation des besoins en eau d'irrigation, evapotranspiration potentielle, formule climatique simplifiée ET mise a jour. (in French). *Annals of Agronomy*, 12, 13–49.

Williams, L.E. and Ayars, J.E. (2005). Grapevine water use and the crop coefficient are linear functions of the shaded area measured beneath the canopy. *Agricultural and Forest Meteorology*, 132, 201–211.

Wright, J.L. (1985). Evapotranspiration and irrigation water requirements. In advances in evapotranspiration. *Proceedings National Conference Advances Evapotranspiration*. American Society of Agricultural Engineers, Chicago, IL. pp. 105–113.

https://doi.org/10.1142/9789811296062_0014

Chapter 14

Modern Concept in Water Management under Changing Climatic Scenario

Vinod Kumar[*,‡] and Ratnesh Kumar Jha[†]

*Department of Agronomy,
RPCAU, Pusa, Samastipur, Bihar, India
†Centre for Advance Studies on Climate Change,
RPCAU, Pusa, Samastipur, Bihar, India
‡vinod_pusa@yahoo.com

Abstract

Water resources are becoming extremely scarce under changing climatic conditions. For judicious use of water, it is pertinent to use proper method, time, and frequency of irrigation as per the requirement of crops. In India, the surface method of irrigation is commonly practiced, which has a very low irrigation efficiency. Pressurized irrigation system is a better system for increasing productivity per unit volume of water besides saving water and other input cost. Drip fertigation in fruits and vegetables produces higher yields and better quality with lesser quantity of water and fertilizers. Drip and sprinkler irrigation also proved to be better options for field crops. Proper land leveling is also required for higher water and nutrient use efficiency. Scheduling of irrigation in crops is normally done on the basis of soil moisture and physiological growth stages. However, climatic parameters play a predominant role in governing the needs of the crop. Thus, the depth of irrigation/cumulative pan evaporation ratio is used for scheduling irrigation.

Keywords: Pressurized irrigation system, Drip fertigation, Land leveling, Climatic parameters, Irrigation scheduling.

Introduction

About 97% of Earth's water is in the sea and only about 1% is readily available out of 3% fresh water. Out of 1% fresh water, about 70% is used for irrigation in agriculture, 22% is consumed by industry, and 8% is used for household purposes. Water resources are becoming extremely scarce (Falkenmark and Rockstrom, 2006; Gleick, 2003; Samakhtin et al., 2004; Turrel et al., 2011). The average annual rainfall of India is about 1,200 mm as against the world's average of 1,100 mm. India is a victim of uneven distribution of rainfall. In fact, one-third of the country is always under threat of drought. Currently, only 29% of the total precipitation is conserved, and that too is not optimally utilized.

Irrigation water which is also known as "Liquid gold" is the most vital and critical input for agricultural production. Out of 140 million hectares of net cultivable area of the country, 68 million hectares of land are under irrigation, which comes to 48% of the net cultivable area, rest of the 52% area is rain-fed. Proper management of these irrigated areas requires the selection of methods, time, and frequency of irrigation to get the maximum water use efficiency (United Nations, 2018).

Method of Irrigation

The most common method of irrigation practiced in India is surface irrigation with very low irrigation efficiency (below 40%). Available estimates indicate that by 10% increase in water use efficiency, the country can gain more than 50 million tons of food grains from the existing irrigated area. Consequently, it will become imperative to adopt modern irrigation technologies to increase water use efficiency. It is critical to employ the correct method to minimize the adverse effects of irrigation. The method of irrigation is influenced by the soil type, land topography, crops to be grown, quality, and quantity of water available for irrigation.

(a) *Pressurized irrigation system*: Pressurized irrigation system is a versatile management tool that can increase productivity per unit volume of water and also save up to 50% of water in addition to other savings in farm input cost (Molden et al., 2010). However, its initial capital cost is high. Without the Government's active financial support and incentives, it is unlikely to become popularly acceptable. It may be a better system for intensive cultivation of fruits, vegetables, and rain-fed areas. Drip fertigation in fruit and vegetable crops produced higher yields and better quality with lesser quantity of water and fertilizers. Besides fruit and vegetable crops drip and sprinkler irrigation in field crops also proved a better option.

(b) *Wheat on sprinkler*: At Prabhani, wheat irrigated through sprinklers with 5 cm depth at 5 critical growth stages (CRI, tillering, late jointing, boot/flowering, and milk stages) produced 5.13 t/ha which was 17% higher than surface irrigation. At Hisar, in sandy loam soil, irrigation with mini-sprinklers produced a higher grain yield of 5.67 t/ha which was 19% higher than surface method and saved 29% water.

(c) *Maize on drip*: At Madurai, in sandy clay soil, hybrid maize recorded the highest grain yield of 8.32 t/ha with 125% recommended dose applied as drip fertigation with fully water-soluble fertilizers and resulted in 18% savings in water and 104% increase in water use efficiency compared to surface irrigation (furrows) and normal soil application of fertilizers.

(d) *Sugarcane on drip*: In sandy loam soil of Hisar, drip irrigation in sugarcane resulted in 11.2% higher cane yield compared to furrow irrigation. Powerkheda and Rahuri also recorded the same result.

(e) *Land leveling*: Proper land leveling is one of the management options, which is not given the importance it deserves. It increases the water application efficiency, which leads to higher yields, more uniform crop growth, as well as increases in water use efficiency. Laser leveling is now gaining popularity as fields can be leveled precisely with this technique. Land leveling has a direct impact on nutrient use efficiency.

Recent Advances in Irrigation Scheduling

"When and how much to irrigate" or scheduling of irrigation can be assessed based on the following three approaches:

(i) Soil moisture as a guide
(ii) Plant as a guide
(iii) Climate as a guide

(i) *Soil moisture as a guide*: Scheduling of irrigation in crops can be done through the soil moisture concept. The basis of assessment may be the traditional feel and appearance of soil (mud ball), percent soil moisture content, soil moisture tension, and depletion in available soil moisture. According to depletion in soil moisture concept, the water content at field capacity was considered 100% available for crop growth, and that at PWP as 0% available. The safe limit of available soil moisture depletion for a crop may be 40%, 50%, or 60% determined on the basis of the experiment. Various instruments like tensiometer, gypsum block, neutron moisture meter, and time domain reflectometer (TDR) may be used for quick measurement of soil moisture.

(ii) *Plant as a guide*: Scheduling of irrigation can be made by indicator plants, leaf water potential, relative water content, and physiological growth stages. Some crucial stages of growth were found to be more critical in their demands for water than others. Indicator plants like sunflower can be used easily for irrigation scheduling. Leaf water potential is likely to be in the region of −2 to −8 bar when plants are in soil with sufficient moisture. Leaf water potential is measured with Thermocouple Psychrometer or Pressure Chamber. Relative water content can be calculated by the following formula:

$$\text{RWC} = \frac{\text{FW} - \text{DW}}{\text{TW} - \text{DW}} \times 100$$

where, FW = Fresh weight of leaf
TW = Turgid weight of leaf
DW = Dry weight of leaf

(iii) *Climate as a guide*: Though most of the data on the scheduling of irrigation is based on either soil water or plant water status as a guiding factor for scheduling irrigation, it is recently realized that climatic parameters play a predominant role in governing the needs of crops (Wada *et al.*, 2011). This led to the concept of evapotranspiration and pan evaporation as the basis for scheduling irrigation. Nowadays, the depth of irrigation water/cumulative pan evaporation (IW/CPE) ratio is used for scheduling irrigation. The criteria of soil water availability or plant water status for scheduling irrigation cannot be considered in isolation from that of climatic factors. Therefore, the recent concept is based on soil plant atmospheric continuum.

Conclusion

Drip and sprinkler irrigation should also be adopted in field crops besides fruits and vegetable crops for higher yields and quality with lesser quantity of water and fertilizers. Proper land leveling should be done for higher water and nutrient use efficiency. Scheduling of irrigation should be followed on the basis of soil water and atmosphere continuum.

References

Falkenmark, M. and Rockström, J. (2006). The new blue and green water paradigm: Breaking new ground for water resources planning and management. *Journal of Water Resources Planning and Management*, 132(3), 129–132.

Gleick, P.H. (2003). Global freshwater resources: Soft-path solutions for the 21st century. *Science*, 302(5650), 1524–1528.

Molden, D., Oweis, T., Steduto, P., Bindraban, P., Hanjra, M.A., and Kijne, J. (2010). Improving agricultural water productivity: Between optimism and caution. *Agricultural Water Management*, 97(4), 528–535.

Rockström, J., Karlberg, L., Wani, S. P., Barron, J., Hatibu, N., Oweis, T., ... Farahani, J. (2010). Managing water in rainfed agriculture — The need for a paradigm shift. *Agricultural Water Management*, 97(4), 543–550.

Smakhtin, V., Revenga, C., and Döll, P. (2004). A pilot global assessment of environmental water requirements and scarcity. *Water International*, 29(3), 307–317.

Turral, H., Burke, J., and Faures, J. M. (2011). *Climate Change, Water and Food Security*. Food and Agriculture Organization of the United Nations (FAO).

United Nations. (2018). *Sustainable Development Goal 6: Synthesis Report on Water and Sanitation*. UN.

van Oel, P.R., and de Fraiture, C. (2018). Water scarcity footprints: Essential to water management in an era of globalization. *Water Resources Management*, 32(3), 849–865.

Wada, Y., van Beek, L.P.H., Viviroli, D., Dürr, H.H., Weingartner, R., and Bierkens, M.F.P. (2011). Global Monthly Water stress: 1. Water balance and water availability. *Water Resources Research*, 47(7), W07517.

Zhang, X., Zhao, Y., Chen, X., and Yin, Z. (2011). Adapting to climate change: Water management for sustainable agriculture. *Agricultural Water Management*, 98(11), 1742–1749.

Chapter 15

Water Management in Different Crops under Changing Climatic Conditions

Vinod Kumar[*,‡] and Abdus Sattar[†]

*Department of Agronomy, RPCAU, Pusa, Samastipur, Bihar
†Agrometeorology, Centre for Advance Studies on Climate Change,
RPCAU, Pusa, Samastipur, Bihar, India
‡vinod_pusa@yahoo.com

Abstract

Check basin method of irrigation is suitable for rice crops to control run-off loss. The practice of intermittent and shallow submergence of 5 ± 2 cm depth during critical growth stages, i.e., tillering, panicle initiation, flowering, and grain filling stages and maintenance of saturation to field capacity during the rest of the stages are beneficial for saving water. About 7 cm depth of each irrigation at 3–7 days after the disappearance of ponded water based on soil and climatic conditions is found to be economical. The optimum soil moisture depletion is required to be kept around 50% in 0–60 cm soil depth, and on the basis of climatological parameters irrigation is required to be given at IW/CPE ratio of 1.0 for wheat and maize. Physiological growth stages are the most common basis for both crops. Although CRI, jointing, booting or flowering, and milky stages are important, CRI is the most critical stage for irrigation in wheat. In maize, knee height, tasseling, silking, grain formation, and dough stages are critical stages, while tasseling and silking stages are the most crucial for irrigation. Border method of irrigation is the most common for wheat. Around 6 cm depth of each irrigation is to be given up to 70% length of the border for saving water. In order to save water alternate furrow method is beneficial for maize. About five irrigations are required from the germination to the tillering stage in sugarcane. At least one irrigation is required at the flower initiation stage for mustard and pulse crops.

Keywords: Check basin method, Intermittent and shallow submergence, IW/CPE ratio, Border method, Alternate furrow method.

Introduction

Increasing water scarcity with variations in climatic conditions threatens the production of different crops. Rice is a major consumer of water resources and requires proper water management to increase water use efficiency. Water requirement of rice varies between 80 and 120 cm depending upon the maturity of its varieties, season, soil type, and climatic conditions. Irrigation scheduling is done on the basis of soil water status, leaf water potential, physiological growth stages, and climatological parameters in wheat, maize, and other aerobic crops. Although physiological growth stages are the most common basis for irrigation, climatological approach may be most suitable for higher productivity and water saving (Hatfield *et al.*, 2011; Kang *et al.*, 2009; Lobell *et al.*, 2011).

Rice

Rice occupies about 50% of irrigated area in India and rice alone consumes about 65% of the total volume of water used in agriculture. It is a semi-aquatic plant, hence its water requirement is many times more than most other food crops. It is a major consumer of water resources of the country and needs careful water management to increase its water use efficiency.

(a) *Water requirement*: Total water requirement of rice varies between 80 and 120 cm depending upon the maturity of its varieties, season, soil type, and climatic conditions. Out of the total water supply only 30–32% is utilized by the plants and the rest is lost by deep percolation and run-off.

(b) *Method of irrigation*: Generally, check basin method of irrigation is adopted for rice crops.

(c) *Depth and duration of land submergence*: Experiments conducted at various locations in India on medium land rice soil reveal that the practice of keeping the soil under shallow depth of submergence $(5 \pm 2\,\text{cm})$ throughout the growth period is conducive to higher yields. The practice of shallow submergence directly saves a considerable amount of water as compared to deep submergence $(10 \pm 2\,\text{cm})$. These results in general hold good under different climatic conditions. The practice of continuous shallow submergence, however, is possible only when the water supplied is adequate and assured. Land also needs to be well-leveled to facilitate the uniform spreading of water.

(d) *Intermittent submergence*: In *kharif* season, when the humidity is high and the evaporative demands are low, it is not necessary to keep the land under continuous shallow submergence. Under these conditions, intermittent submergence, i.e., submergence during critical growth stages (tillering, panicle initiation, flowering, and grain filling stages) and maintenance of saturation to field capacity during the rest of the stages resulted in a similar yield as in the case of continuous submergence. Intermittent submergence (AWD) has been found beneficial in economizing water and allowing soil for a free diffusion of oxygen in the system. Intermittent submergence further saved 30–50% of water in comparison to shallow submergence of water. Based on the 3 years experiment at Pusa, it was observed that 0–20 and 40–60 days after transplanting which coincides with tillering and panicle initiation were more critical in case of early rice varieties than other stages with respect to irrigation. If there was water stress during these periods the yield reduction went to 20–30% over continuous ponding. In the Gandak command area and in the rest of Bihar the general recent recommendation for scheduling irrigation in rice is to apply 7 cm irrigation 3 days after the disappearance of ponded water (Choudhary *et al.*, 2012). However, in Sone command the days of disappearance should be 5 days due to clay soil and in Koshi command, the days of disappearance may be as high as 6–7 days because of high water table.

(e) *Water and weed management under SRI*: Among the various factors of production, water and weed management practices are of prime importance. In SRI, soil is to be kept both moist and aerated, at least during the vegetative growth period. Mechanical weeding by rotary weeder starting 10 days after transplanting and repeated 2–3 times at 10–12 days interval churns the soil and incorporates the weeds into the soil (Kumar *et al.*, 2015).

Wheat

Irrigation is scheduled in wheat on the basis of soil moisture depletion, soil moisture tension, leaf water potential, physiological growth stages, and climatological approach (Allen *et al.*, 1998; Asseng *et al.*, 2015). The optimum soil moisture depletion for wheat should be around 50% in 0–60 cm soil depth, while in case of tension it may be around −0.5 bar in the same layer. Under these conditions, the total number of irrigation varies according to the soil and climatic conditions of different places. However, a better approach based on climatological parameters is now in vogue which is termed as IW/CPE ratio which should be 1.0. For the benefit of farmers this approach can be easily translated in terms of days after sowing and can be seen from the calendar developed for wheat crop.

(a) *Critical stages of water need for wheat crop*: Physiological growth stages in wheat are the most common basis for irrigation scheduling. Among the different physiological stages, the CRI stage (20–25 days after sowing) is the most crucial for irrigation, which results in maximum production per unit of water applied. The second most critical stage of irrigation is the flowering stage. The recent view is that under limited water supply, the boot stage is the second most important stage in comparison to flowering. Among other stages jointing or milk ranks third followed by the dough stage (Table 1). First irrigation at CRI should be light and

Table 1.   Irrigation on a priority basis.

No. of irrigations available	Physiological growth stages when water is applied
One	CRI
Two	CRI and Boot
Three	CRI, Jointing, and Flowering
Four	CRI, Jointing, Flowering, and Milk
Five	CRI, Jointing, Flowering, Milk, and Dough

subsequent irrigation should be 6 cm each. If the water table is high the depth of irrigation may be reduced.

(b) *Irrigation scheduling on the basis of climatological approach*: Experiments conducted for four years at Pusa have shown that wheat responded well up to 4 irrigations based on IW/CPE ratio of 1.0. The first irrigation was given at the CRI stage and the remaining irrigation fell on 65, 90, and 110 days after sowing according to the date of sowing, evaporative demand, and rainfall pattern. The experiments have shown that the response of wheat to irrigation was dependent on the level of nitrogen application. Under limited irrigation, the response to nitrogen was well marked up to 80 kg/ha. It was also observed that if the nitrogen dose was increased to 120 kg/ha, there was a decrease in yield. At optimum irrigation, the response to nitrogen was noticed up to 100 kg/ha. At higher levels of irrigation, the response to nitrogen was noticed up to 120 kg/ha.

(c) *Irrigation calendar for wheat*: Based on experiments conducted using IW/CPE ratio approach, an irrigation calendar for wheat crop has been developed for the Gandak Command Area for different dates of sowing (Table 2). The fourth irrigation for normal sown and the third for late sown wheat will invariably not be needed except under adverse weather conditions. In case of rainfall during the crop period, the succeeding irrigation, for each centimeter of rain, may be delayed by 5, 6, 4, and 2 days, respectively in the months of December, January, February, and March. However, rainfall below 5 mm may not be considered effective for this purpose.

Table 2. Irrigation calendar for wheat.

Date of sowing	Irrigation			
	I	II	III	IV
8–15 Nov	21	60	93	114
16–22 Nov	21	60	85	110
23–30 Nov	21	61	90	105
1–7 Dec	25	63	88	103
8–15 Dec	25	62	85	100
16–22 Dec	25	62	83	—
23–31 Dec	25	60	80	—

(d) *Method of irrigation in wheat*: The wheat crop is irrigated mostly
    by the border method of irrigation. This is a controlled method of
    surface irrigation and more efficient than flood irrigation. In case
    of border irrigation, the field is divided into a number of strips
    separated by small bunds and water is allowed in each strip sepa-
    rately. The border strip should have a mild longitudinal slope and
    a mostly leveled surface throughout the width. Water is allowed
    at the upstream which moves downward due to longitudinal gra-
    dient irrigating the crops. The general practice of allowing the
    irrigation water up to the end of the border of fields results in
    excess irrigation in most cases and water stagnation at the tail-
    end. Such practice also leads to loss of nutrients and thereby
    reduction in yield of crops. Therefore, water supply in individual
    borders should be stopped as soon as water approaches 70% of
    the length of the border. The remaining length of the border is
    irrigated due to recession of surface water. The depth of irriga-
    tion is uniform under 70% cut-off and varies from 5 to 7 cm for
    50 m long borders. There is 20–30% saving of water at 70% cut-
    off as compared to 90% cut-off. For efficient irrigation in wheat
    crop, the design parameter of the border for wheat crops is given
    as follows:

- Border length: 40–50 m
- Border width: 3–5 m
- Slope: 0.2–0.3%
- Inflow rate per unit width: 2 lps/m
- Cut-off distance: 70%

Wheat should be irrigated by sprinkler method for higher yield
and water use efficiency.

Maize

Maize is grown in all the seasons in Bihar. Previously it was grown
predominantly in the *kharif* season but nowadays, its cultivation is
done in *rabi* and *zaid* seasons. The productivity of only maize crops
is the highest in Bihar than national productivity. This is mainly

due to the cultivation of maize in *rabi* season. Irrigation is scheduled in maize on the basis of soil moisture depletion, soil moisture tension, leaf water potential, physiological growth stages, and climatological approach. The optimum soil moisture depletion for maize should be around 50% in 0–60 cm soil depth, while in case of tension it may be around −0.5 bar in the same layer. Under these conditions, the total number of irrigations varies according to the soil and climatic conditions of different places. However, a better approach based on climatological parameters is now in vogue which is termed as IW/CPE ratio. For the benefit of the farmers this approach can be easily translated in terms of days after sowing.

(a) **Critical stages of water need for maize crop:** Physiological growth stages in maize are the most common basis for irrigation scheduling. Among the different physiological stages in maize the crucial growth stages are knee high, tasseling and silking, grain development, and dough stages. At these stages there should not be water stress for proper growth of plants, otherwise there will be much reduction in yield as compared to stress at other stages. Tasseling and silking and knee-high stages are the most and next important stages, respectively, for irrigation in maize.

(i) *Kharif crop*: *Kharif* maize generally does not require additional irrigation but when the rain fails at an important growth stage, 1–2 supplemental irrigation is required. During this season proper management of excess water through appropriate drainage system is more important. It has been observed that poor drainage or waterlogging for 3–4 days may reduce maize yield by 80–90%.

(ii) *Rabi crop*: In winter maize 5–6 irrigations each of 6 cm depth at IW/CPE ratio of 1.0 has been found beneficial (Amandu *et al.*, 2015). On the basis of trials, the first irrigation falls between 35 and 40 DAS, and the subsequent 2nd, 3rd, 4th, 5th, and 6th irrigation fall at 50–55, 75–80, 95–100, 110–115, and 135–140 DAS, respectively.

(iii) *Zaid crop*: Experiment conducted on *zaid* crop reveal that 5–6 irrigations are required scheduled at 1.0 IW/CPE ratio. On the basis of trials, the first irrigation should be given at 30 days after sowing (DAS), and the subsequent 2nd, 3rd,

4th, 5th, and 6th irrigation at 40, 50–55, 63–65, 73–75, and 82–85 DAS, respectively. Response of nitrogen was up to 150 kg/ha.

(b) **Method of irrigation in maize:** In maize, farmers mostly use the furrow method of irrigation. This method is also scientifically sound for the crop. In order to save water this approach has been slightly modified at Pusa. In place of allowing water in each furrow, irrigation water is allowed in alternate furrows. In place of alternate furrow irrigation, skip furrow irrigation may be equally effective. In case of skip furrow, furrows are opened at a distance equal to two furrows or furrows are opened after two lines of maize. This also saves money while calculating the total cost of cultivation.

It is seen from Table 3 that by deviating from the normal practice of irrigation in each furrow, a considerable amount of water can be saved from other methods. Although in paired row, about 23.1% of water can be saved, the yield is low and thereby the water use efficiency is at par with irrigation in each furrow. However, alternate or skip furrow methods help to save irrigation water about 23.9% and 28.4%, respectively, without any appreciable reduction in grain yield. Therefore, in case of a deficit irrigation water supply, either of these methods could be adopted.

Table 3. Effect of different irrigation methods on grain yield, water-use-efficiency, depth of irrigation, and percentage saving of water in maize.

S. No.	Irrigation methods	Grain yield (q/ha)	Total depth of irrigation (cm)	WUE (q/ha-cm)	Percentage saving in water
1.	Irrigation in each furrow	43.6	26.8	1.626	—
2.	Irrigation in alternate furrow	40.9	20.4	2.002	23.9
3.	Irrigation in paired furrow	38.3	20.6	1.858	23.1
4.	Irrigation in skip furrow	39.3	19.2	2.044	28.4

Nowadays, furrow irrigated raised bed system (FIRBS) is adopted in mechanized cultivation. Yield is higher due to good aeration, land leveling, and uniform size of furrows. The space of the furrow varies from 50 to 100 cm depending upon the type of soil. For sandy soils, it should be 50 cm while for sandy loam around 60–70 cm. For clay soils, 80–100 cm furrow spacing is preferred. The length of the furrow shall depend on the type of the soil as well as the size of the stream.

(c) **Rabi maize + potato intercrops:** Rabi maize + potato intercrops are the most beneficial intercropping system for north Bihar agro-climatic conditions and in this system highest yield and net profit are obtained with 5–6 irrigations at 1.0 IW/CPE ratio (Kumari *et al.*, 2015). Application of 100% recommended dose of fertilizer (120 kgN75 kgP_2O_5 and 50 kgk_2O/ha.) in maize with 75% recommended dose of 150 kgN90 kgP_2O_5 and 100 kgK_2O/ha in potato at this level of irrigation is most profitable.

Sugarcane

In winter-planted sugarcane 5–6 irrigations are required while spring-planted sugarcane only 4–5 irrigations are required at the tillering stage at 20-day intervals. Water is also required at the grand growth stage which falls in the rainy season. If rain fails irrigation should be given. Irrigation should be given in alternate furrow which saves about 36% water in comparison to irrigation in each furrow.

Other Crops

Except the water needs of rice, wheat, maize, and sugarcane, the irrigation requirement of other major crops is mentioned in Table 4.

Conclusion

Intermittent and shallow submergence during critical growth stages should be maintained for a higher yield of rice and water saving. Irrigation should be given at IW/CPE ratio of 1.0 for wheat and maize for higher production and water use efficiency. There should not be water stress at the CRI stage in wheat and tasseling and

Table 4.   Summary of irrigation requirement of major crops.

Crop	Total no. of irrigations	Time of application		Total irrigation requirement (cm)
		Physiological stage	DAS	
Barley	2	Tillering	30	12
		Flowering	80	
Mustard	2	Flower initiation	30	12
		Pod formation	60	
Gram	1	Flower initiation	50	5
Pea	2	Flower initiation	45	10
		Pod formation	60	
Lentil	1	Flower initiation	60	10
Rajmash	2	Grand growth stage	25	12
		Pre flowering	50	
Moong	1–2	Pre sowing (if needed)	—	6–12
(Summer)		Flower initiation	30	
Potato	3–4	Each irrigation at an interval of 20 days	—	18–24
Tobacco	2	Ball formation	25	12
		After final topping & piercing	50	

silking stages in maize. Irrigation should be applied at critical growth stages in other aerobic crops.

References

Allen, R.G., Pereira, L.S., Raes, D., and Smith, M. (1998). Crop evapo-transpiration: Guidelines for computing crop water requirements. *FAO Irrigation and Drainage Paper 56*. Food and Agriculture Organization of the United Nations, Rome.

Amandu, L.P.M., Kumar, V., Kumar, M. and Kumar, R. (2015). Influence of mulching and irrigation regimes on yield and water-use efficiency of maize-maize cropping system. *RAU Journal of Research*, 25(1&2), 5–8.

Asseng, S., Ewert, F., Martre, P., Rötter, R. P., Lobell, D. B., Cammarano, D., ... Rosenzweig, C. (2015). Rising temperatures reduce global wheat production. *Nature Climate Change*, 5(2), 143–147.

Choudhary, D.K., Kumar, V., Bharati, V. and Kumar, S.B. (2012). Effect of water regimes and NPK levels on yield, quality assessment and water-use-efficiency on mid duration rice. *Journal of Interacademicia*, 16(3), 639–642.

Hatfield, J.L., Boote, K.J., Kimball, B.A., Ziska, L.H., Izaurralde, R.C., Ort, D., and Thomson, A.M. (2011). Climate impacts on agriculture: Implications for crop production. *Agronomy Journal*, 103(2), 351–370.

Kang, Y., Khan, S., Ma, X., and Zhang, J. (2009). Climate change impacts on crop yield, crop water productivity and food security — A review. *Progress in Natural Science*, 19(12), 1665–1674.

Kumar, R., Das, S., Kumar, V., Dwivedi, D.K. and Das, L. (2015). Studies on irrigation and weed management for enhancing rice yield and water productivity under system of rice intensification. *The Bioscan*, 10(1), 417–420.

Kumari, R., Nandan, R. and Kumar, V. (2015). Effect of crop diversification and moisture regimes on productivity and water use efficiency under rice-based cropping system. *New Agriculturist*, 26(1), 59–63.

Lobell, D.B., Schlenker, W., and Costa-Roberts, J. (2011). Climate trends and global crop production since 1980. *Science*, 333(6042), 616–620.

Index